THE REALITY
OF GOD

THE REALITY OF GOD

THE LAYMAN'S GUIDE TO SCIENTIFIC EVIDENCE FOR THE CREATOR

Steven R. Hemler

Saint Benedict Press
Charlotte, North Carolina

Nihil Obstat: Rev. Paul deLadurantaye, S.T.D.
                    Censor Librorum

Imprimatur: ☩ Paul S. Loverde
                    Bishop of Arlington
                    January 14, 2014

Cataloging-in-Publication data on file with the Library of Congress

Cover design by David Ferris
www.davidferrisdesign.com

Cover image: Getty Images / Marco Lorenzi,
www.glitteringlights.com

ISBN: 978-1-61890-713-4

Published in the United States by
Saint Benedict Press, LLC
PO Box 410487
Charlotte, NC 28241
www.SaintBenedictPress.com

Printed and bound in the United States of America

CONTENTS

Part III: Human Evidence of God's Existence

God Is . . .

ACKNOWLEDGMENTS

I gratefully acknowledge the many contributions of family and friends in the development of this book. I am very grateful for the review comments and excellent suggestions provided by Fr. Peter Reynierse, Fr. Paul Koenig, Dr. Anthony Foster, Susan and Mark Hope, Jill Devine, Bob Wiersberg, Jon Brick, Billie Marie Springart, Scott Hemler, Dan Marcum, Randy Hyde, Dave Hrivnak, and several others. I am also very appreciative of Sr. Mary Margaret Ann Schlather and Carol Daley of Catholic Distance University for implementing a new three-week apologetics seminar on evidence for God's existence, as it was this online seminar that prompted me to adapt several of our apologetics presentations into the text-based lecture format that was the basis of this book. The help of R. C. and Marc Garcia in developing the initial draft of these seminar lectures was invaluable. Of course, I am especially grateful for the love and support of my wife, Linda, and our children, Jonathan, Christopher, and Allison. I have been truly blessed with a wonderful wife and fantastic children. May God bless everyone involved in this labor of love and all those who read this book, which was prepared to help manifest God's majesty as revealed in His creation.

PREFACE

The question "Does God exist?" is critical to understanding our place in the universe and how we will live our lives. That's because what someone believes about God affects everything else he or she believes and does. Indeed, the five most consequential questions in life are these:

1. Origin: Where did we come from?
2. Identity: Who are we?
3. Meaning: Why are we here?
4. Morality: How should we live?
5. Destiny: Where are we going?"

Our answer to each of these questions and consequently how we live our lives depends upon whether or not we believe God exists. Therefore, this book provides a **basic introductory overview** of evidence for the existence of God to help us answer the foundational question, "Does God exist?" for ourselves and in discussions with others. As such, this book is intended to help strengthen our own faith and provide valuable information for discussion with those who may doubt the existence of God.

External and Internal Challenges to Faith

There are major challenges to faith in today's increasingly secular Western culture. Both **external** challenges (from outside ourselves) and **internal** challenges (from within ourselves) to belief in God have caused many people to stop practicing the faith of their childhood.

According to Pew Research Center surveys, nearly a quarter (23 percent) of American adults say they have no religious affiliation, an increase from 16 percent in just seven years.[1] Furthermore, over one-third of young adults under age 30 have no religious affiliation, the highest percentage ever in Pew Research Center polling. Pew's surveys also find that men are more likely than women to claim no religious affiliation. More than a quarter of men (27 percent) say they have no religious affiliation, compared with 19 percent of women. Finally, the Pew Forum's U.S. Religious Landscape Survey found that fully 41 percent of those who were raised Catholic no longer identify themselves as Catholic.[2]

These trends are not surprising given that our increasingly secular society often does not support or allow much room for religious beliefs or convictions in the public sphere. It is especially difficult for a young person to maintain his or her faith when faced with so much that is contradictory to that faith from the media, including much of what we find in movies, television, music, video games, and the internet. Distinguishing truth from falsehood in our relativistic culture is increasingly difficult, since trendsetters mostly speak and act as if God and objective truth do not exist.

Another external challenge to faith is intellectual in nature. The general world view within the scientific and academic communities is that the material world of nature, that which can be observed, measured, and quantified by science, encompasses the whole of reality. This philosophical world view that nature is "all there is" is called materialism or naturalism. Naturalism, however, leaves no room for so-called religious "superstition" or "unsubstantiated" faith. Many people lose their faith when they become convinced that science has all the answers and everything can be fully explained by natural causes only.

Of course, there are scientists and academics who do believe in God. However, the longstanding atheistic leaning of the scientific "establishment" has been repeatedly confirmed by surveys of religious beliefs among scientists. For example, a 1998 survey of leading scientists from the National Academy of Sciences (NAS) found among them:

> . . . near universal rejection of the transcendent by NAS natural scientists. Disbelief in God and immortality among NAS biological scientists was 65.2% and 69.0%, respectively, and among NAS physical scientists it was 79.0% and 76.3%. Most of the rest were agnostics on both issues, with few believers.[3]

They also found, "Biological scientists had the lowest rate of belief (5.5% in God, 7.1% in immortality)." It is even more alarming today with recent surveys showing that well over 90% of the members of the National

Academy of Sciences (the world's top scientists) are atheists. And, one final piece of anecdotal evidence: How often do we see a documentary on television that acknowledges God may have had anything to do with the creation of the universe or the evolution of life? In fact, the recent *Cosmos* television series with Neil DeGrasse Tyson went to great pains to push God to the fringe (at best) in favor of science. Given all this, it is no surprise that many young people leave the faith.

In addition to these and other external challenges, people who are raised in faith may face internal doubts about their beliefs as they grow older. These internal challenges are best understood by looking at what has been called the "stages of faith."

In his classic book, *Will Our Children Have Faith?*, John Westerhoff describes four stages of Christian faith development:[4]

1. Experienced Faith
2. Affiliative Faith
3. Searching Faith
4. Owned Faith

According to Westerhoff, as we go through life we grow into (add on) the needs and elements of the later stages, but only if the needs of the earlier stages are met. The stages of faith are like rings of a tree. Just like a tree adds one ring on top of another, we do not leave the needs of the earlier stages behind after we expand into the next stage. We continue to need the faith experiences of the earlier stages throughout our lives.

Experienced Faith grows by participating in

(experiencing) the customs and rituals of one's faith tradition (the "bells and smells"). It is first experienced by young children and is the lifelong foundation of faith.

Affiliative Faith develops by belonging to (being affiliated with) a close Christian community where one is individually known, valued, and accepted. Provided the needs of experienced faith are met during childhood, young people may expand into the needs of affiliative faith during adolescence. A dynamic youth ministry program, with its social, spiritual, and service activities, provides excellent opportunities for young people to deepen their relationships with Christian peers. However according to Westerhoff, research shows that the faith growth of most adults has been "arrested" at the affiliative faith stage.[5]

Searching Faith recognizes and explores personal questions and doubts about one's faith. Searching faith is the faith of questioning and internalizing what we have long been taught. Searching faith usually begins during late adolescence and often continues in earnest during young adulthood. College students frequently have discussions on the existence of God, evolution, Jesus' divinity, etc., with their professors and fellow students. This "Searching Faith" stage can be troubling if not properly understood. And, of course, it is risky. However, only by questioning and testing what we have long been taught can we truly come to personally accept and internalize these teachings. This book is particularly relevant in helping dispel doubts about God's existence that typically arise during the "Searching Faith" stage of faith development.

New York Cardinal Timothy M. Dolan spoke on the importance of asking questions and facing our doubts at the World Youth Day in Madrid in August 2011 when he said, "When we admit our faith is weak, when we admit our faith is shaky, when we admit that our faith isn't what it should be, actually we're exercising it, and we're making it more and more firm."[6]

Owned Faith rarely occurs before young adulthood. Because of the serious struggle with doubt that precedes it, owned faith may appear as a great illumination or enlightenment. It is now one's own faith and not merely the faith of one's family or friends. People who "own" their faith choose to be involved in religious activities because they want to, and not just because they are expected to. Even though some doubts and questions may remain, those who "own" their faith seek to witness it by personal and social action, and are willing and able to stand up for what they believe as mature disciples of Jesus Christ. According to Westerhoff, owned faith is God's intention for everyone. However, the only path to owned faith is to work through the doubts and questions of the searching faith stage.

This book is intended to help us face these external and internal challenges to faith in God. As such, it is meant to help us and those we love better respond to external challenges to faith, as well as address our own searching faith doubts and questions. Thus, the goals of this book are to *strengthen* our faith in God and to enable us to *better explain* reasons for belief in God's existence.

This Book Provides Objective Reasons for Belief in God's Existence

While we may have subjective (personal) reasons for belief in God, this book provides objective (scientific and philosophical) reasons that justify God's existence. As such, this unique book provides a clear and concise overview of key scientific evidence, philosophical reasons, and insights drawn from human nature itself demonstrating God's existence.

In this book, you will find "the rest of the story" (to quote famous radio personality Paul Harvey). In the first two parts of this book, we explore aspects of scientific discoveries that you are unlikely to hear in scientific lectures, publications, or television shows; namely, how recent scientific discoveries actually provide compelling evidence of God's existence.

Thus, this is a book of Christian apologetics, which is the branch of theology that is concerned with the rational explanation and defense of the Christian faith.[7] Cardinal Timothy Dolan, while he was President of the United States Conference of Catholic Bishops (USCCB), spoke on the importance of apologetics today when he stated during a September 2013 speech in Milwaukee,

> We have failed to equip our young people and ourselves with the art of credibly, convincingly, and compellingly defending and presenting our beloved faith. . . . So, we need a renewed sense of apologetics . . . I mean a humble, cheerful, confident, rational grounding in our Catholic faith . . .

Apologetics prepares us to survive a rational criti-
cism of our faith that is all over the place today. We
need it more than ever.[8]

And, Pope Francis in his apostolic exhortation *The
Joy of the Gospel* (*Evangelii Gaudium*) states,

Proclaiming the Gospel message also involves proclaiming it to professional, scientific and academic
circles. This means an encounter between faith,
reason and the sciences with a view to developing new approaches and . . . a creative apologetics
which would encourage greater openness to the
Gospel on the part of all.[9]

As such, the first part of this book explores the body
of scientific evidence of God's existence found in the
cosmos. This evidence includes the Big Bang, the elegant laws of nature, and the "fine tuning" of the universe
and our solar system for life. This evidence, which God
left for science to discover, points to a Creator beyond
space and time who brought the universe into being and
designed it for life.

The second part of this book presents biological
evidence of God's existence found within the cells of all
living things. This part also examines how the theory of
life's evolution over time is indeed compatible with faith
in God. In this part, we initially look at how to properly interpret the Genesis creation stories, followed by
the concept of evolutionary creation. This is contrasted
with evolutionary naturalism, which is the prevailing
view of evolution within academia and the scientific

establishment, and how a "layered explanation" offers a more complete understanding of evolution. We close this second part of the book by looking at how the genetic information in DNA provides biological evidence of God's existence.

In the third part of this book, we move beyond science to present philosophical evidence of God's existence from within human nature and by human reason. This part focuses on evidence within ourselves for God's existence, as well as rational philosophical reasons to believe in the existence of God. This includes examining how human consciousness—including our innate desire for full knowledge and perfect love—and our moral conscience point to God's existence.

Let's start by looking at evidence of God's existence found in the cosmos, most of which scientists have only recently discovered.

PART I

COSMIC EVIDENCE OF GOD'S EXISTENCE

INTRODUCTION TO PART I

In preparation for the rest of this book, it may be helpful to answer the following questions for yourself.

1. If you or someone you know does not believe in God, what are some of the reasons?
2. If you believe in God, what are some reasons for your belief?
3. Which of these reasons are subjective (based on personal experience) and which are objective (based on science or philosophy)?

Again, while we may have **subjective** reasons for belief in God, in this book we are primarily concerned with **objective** evidence of God's existence. We will start by examining scientific evidence of the existence of God.

But, you may ask, why look to science for evidence of God's existence? While the answer may be surprising to some, it is because examining the physical world is one way the Bible and the Church teach that God may be known.

For example, Psalm 19:1–2 declares, "The heavens declare the glory of God, the skies proclaim the work of his hands. Day after day they pour forth speech, night after night they display knowledge." Wisdom 13:5 proclaims, "For from the greatness and beauty of created things comes a corresponding perception of their Creator." And, at the beginning of his letter to the Romans, St. Paul writes, "Ever since the creation of the world, God's invisible attributes of eternal power and divinity have been able to be understood and perceived in what he has made" (Romans 1:20).

The *Catechism of the Catholic Church* (#31) also states,

> Created in God's image and called to know and love him, the person who seeks God discovers certain ways of coming to know him. These are also called proofs for the existence of God, not in the sense of proofs in the natural sciences, but rather in the sense of "converging and convincing arguments," which allow us to attain certainty about the truth. These "ways" of approaching God from creation have a twofold point of departure: the physical world and the human person.

Thus, we will initially look at scientific evidence supporting belief in God found in the physical world. In the first part of this book, we will examine evidence of God's existence from cosmology (the study of the cosmos), specifically the Big Bang theory of the origin of the universe. Then, from physics and astronomy we will explore how the laws of nature point to the existence of God and how these laws and our solar system are "finely tuned" in order for life on Earth to be possible.

CHAPTER 1

THE ORIGIN OF THE BEGINNING

Almost a century ago scientists discovered that galaxies are moving away from each other at fantastic speeds. Because of this, and a great deal of other evidence, scientists now believe that the universe—the very fabric of space-time—is expanding from a cataclysmic explosion that happened about 13.7 billion years ago, which is known as the "Big Bang." In the currently standard theory of the Big Bang, the universe itself began with that explosion, which was the beginning of matter, energy, space, and time itself.

Most of us have a general understanding of the scientific development of the Big Bang theory. But, let's review a few key milestones.[1]

- First, Albert Einstein's General Theory of Relativity implied that the universe was either expanding or contracting and may have had a beginning. These implications were contrary to the prevailing view from ancient times that the universe was static and eternal. Thus, the implications of his theory that the universe may have had a beginning were not initially accepted, not even by Einstein.

- Then, in 1927 Belgian physicist and Catholic priest Georges Lemaître developed a detailed model based on Einstein's equations showing that our universe is expanding. Four years later, he proposed that this expansion began suddenly at a definite point in the past. This became known as the Big Bang theory. However, this theory was initially met with skepticism by most scientists. For example, renowned British astrophysicist Sir Arthur Eddington declared in 1931, "Philosophically, the notion of a beginning of the present order of nature is repugnant to me . . . I should like to find a genuine loophole."[2]

- In 1929, however, American astronomer Edwin Hubble observed galaxy movements using the large telescope at the Mount Wilson Observatory in California. His observations supported the idea of a universe expanding from the Big Bang.

- There have been other scientific confirmations of the Big Bang theory since 1929 (e.g., cosmic microwave background radiation), so that virtually all cosmologists now accept it.

- Scientists also realized Einstein's equations of gravity indicated that at a finite time in the past there was a point where the energy, density, temperature, and curvature of space-time itself were infinite. In the standard Big Bang theory, this point marked the beginning of the universe.

In the standard Big Bang theory, nothing existed before the Big Bang—not even time—so that one should not even talk about "before the Big Bang." In the standard theory, then, the Big Bang was the beginning of our universe. For example, when scientists use the classical laws of physics to determine how the Big Bang started they encounter what is called a "singularity," a point where "the laws of physics seem to break down and a transcendent God is theoretically needed to start the universe."[3]

However, many speculative variants of the standard theory have been proposed and in some of these theories the universe did exist for a period of time before the Big Bang. Nevertheless, there are strong theoretical arguments even in these alternative theories that the universe must have begun a finite time ago, though maybe at a point earlier than the Big Bang.[4] Thus, the weight of scientific evidence and argument is strongly in favor of the universe having had a beginning.

The Kalam Cosmological Argument

Therefore, the Big Bang supports a medieval philosophical argument for the existence of God called the Kalam

Cosmological Argument.[5] The three logical steps of the
Kalam Cosmological Argument are deceptively clear and
simple in their formulation:

1. Premise One–Whatever begins to exist
    has a cause of its existence.
2. Premise Two–The universe began to
    exist.
3. Conclusion–Therefore, the universe has
    a cause of its existence.

Even renowned skeptic David Hume didn't deny the
first premise that whatever begins to exist has a cause.
And, with the Big Bang theory, there is now solid scientific evidence in support of this second premise that the
universe had a beginning. The conclusion of the Kalam
Cosmological Argument follows inexorably from the
first two premises; namely, the universe has a cause of its
existence.

Thanks to scientific discoveries of the last century,
the traditional Kalam Cosmological Argument for the
existence of God has taken on a powerful and persuasive
new force. And, atheist Quentin Smith's contention that
"we came from nothing, by nothing, and for nothing"
seems absurd.[6]

William Lane Craig, in his book *The Kalam Cosmological Argument*, argues that this logical deduction points
to a Creator beyond the universe, a transcendent reality
beyond space and time and therefore non-physical and
immaterial, who created the universe out of nothing and
brought it into being.[7] In short, as a beginning implies a
cause, a cause implies a creator.

The implications of a universe with a beginning, coupled with other discoveries of modern cosmology, have led some scientists to unmistakably theological conclusions. For example:

- British physicist Edmund T. Whitaker wrote, "There is no ground for supposing that matter and energy existed before and were suddenly galvanized into action. It is simpler to postulate creation *ex nihilo*—Divine will constituting Nature from nothingness."[8]
- Allan R. Sandage, one of the world's leading astronomers, said, "We can't understand the universe in any clear way without the supernatural."[9]
- And, world-renowned astrophysicist Robert Jastrow wrote, "the essential elements in the astronomical and biblical accounts of Genesis are the same; the chain of events leading to man commenced suddenly and sharply at a definite moment in time, in a flash of light and energy." He continued, "For the scientist who has lived by his faith in the power of reason, the story ends like a bad dream. He has scaled the mountain of ignorance; he is about to conquer the highest peak; as he pulls himself over the final rock, he is greeted by a band of theologians who have been sitting there for centuries."[10]

The Big Bang suggests that the universe had a beginning. And, current scientific theory strongly supports the conclusion that the universe had a beginning at some point, either at the Big Bang or earlier. This can be seen as pointing to a Creator who transcends space and time and brought the universe into being. For as it says in the Book of Genesis, it was God who in the beginning said, "Let there be light!" Science has shown that the author of Genesis could have been referring to the "light" from an unimaginably explosive and energetic "Big Bang."

As previously mentioned, alternative theories have been proposed in which the Big Bang itself was not the beginning of the universe. For example, our universe could be part of a bouncing cycle (a pre-Big Bang universe collapsed and then began to expand again) or our universe could be just one of many universes in a "multiverse." These theories can be fascinating to consider, but they require advanced mathematical formulas to explain the physics behind them and are beyond the scope of this introductory book.

What's important to note, however, is in 2003 three physicists (Arvind Borde, Alan Guth, and Alexander Vilenkin) developed what is now called the BGV Theorem (also called the BVG Theorem). This theorem shows that any expanding universe that satisfies certain likely conditions must have a beginning.[11] This means any bouncing universe or multiverse (of which our universe may be a part) must have had an ultimate beginning. As, Dr. Vilenkin has declared,

> It is said that an argument is what convinces reasonable men and a proof is what it takes to convince even an unreasonable man. With the proof

now in place, cosmologists can no longer hide. . . .
There is no escape: they have to face the problem of
a cosmic beginning.[12]

And, even if the spontaneous appearance of our universe "from nothing" (actually a "quantum vacuum") can be explained by the laws of quantum mechanics, as asserted by famed Oxford physicist Stephen Hawking in his book *The Grand Design*, this does not answer the question of why anything exists at all. Besides, this assertion merely raises the question of where did these governing laws of nature come from. This is what we will look at next.

CHAPTER 2

THE ARCHITECT OF THE LAWS OF NATURE

After such a violent and explosive beginning at the Big Bang, it seems reasonable to expect that the universe would behave in a random, chaotic, and unpredictable manner. However, our universe is surprisingly well-ordered and predictable, governed by the laws of physics. And, these governing laws of nature provide more evidence of God's existence.

Let's look at a few examples of the laws of nature. First, in Einstein's famous equation, $E=mc^2$, energy ("E") equals mass ("m") times the speed of light ("c") squared. This is an example of an amazing and powerful law of nature that can be expressed with very simple mathematics. Even though the mathematical equation is very simple, this physical law has far-reaching implications for the entire universe, whether it is generating the power of atomic bombs or the energy of the Sun and the stars.

Another example is the Second Law of Motion formulated by Sir Isaac Newton, which states that the force exerted by an object ("F") equals the mass of the object ("m") times the acceleration of the object ("a"). In other words, F=ma. This very simple mathematical expression has profound usefulness in our everyday world; for example, in the design of our car's brakes.

Newton's universal law of gravity, $F = G \, m_1 \, m_2 \, / \, d^2$, is one more example of a simple mathematical expression with powerful ramifications. In this formula, Newton showed that the gravitational force of attraction between two objects ("F") equals the Universal Gravitational Constant ("G") times the mass of the first object ("m_1") times the mass of the second object ("m_2"), divided by the distance between the two objects ("d") squared.

These are just a few examples of amazing and powerful laws of nature that can be expressed with elegant and simple mathematics. Of course, there are many other physical laws that can be expressed mathematically and have profound ramifications in our universe.

The governing laws of nature provide order, harmony, and regularity to the universe. Without these laws, the universe would be nothing but chaos. Science, which seeks to discover and mathematically quantify the laws of nature, is possible only because the universe is orderly, regular, and predictable. It is truly amazing that many of these physical laws can be mathematically expressed so simply and elegantly.

Science has not eliminated the question of **why** the universe is orderly and lawful. Rather, science continues to reveal that the order and lawfulness in our universe runs far deeper and is more impressive than previously thought. If order, in general, calls for an explanation, then the magnificent order in our universe described by the governing laws of nature surely calls for an explanation.

This requires us to ask: **Where did the laws of nature come from?** In other words, what "breathed reality" into the governing laws of physics?[1]

The harmony, order, and elegance found in the governing principles of the universe, based on mathematical concepts that take the greatest efforts of the finest human minds to unlock and understand, must have come from a Mind far greater.[2] How else could the laws of nature have originated? To quote Albert Einstein,

> The scientist is possessed by the sense of universal causation. . . . His religious feeling takes the form of a rapturous amazement at the harmony of natural law, which reveals an intelligence of such superiority that, compared with it, all the systematic thinking and acting of human beings is an utterly insignificant reflection.[3]

It is not only remarkable how well the laws of nature work, it is also remarkable how mathematically elegant they are and that we humans are capable of understanding them. This realization caused Albert Einstein to remark, "I have deep faith that the principles of the universe will be both beautiful and simple."[4] That simplicity is related to the fact that the laws of nature are comprehensible by human beings. In other words, we live in a universe that not only has a beautiful, elegant mathematical structure but one that can be discovered and understood by humans. The spectacular progress of modern science is the result of a universe that is accessible to the human eye and understandable by the human mind.

This caused Albert Einstein to marvel, "The most incomprehensible thing about the universe is that it is comprehensible."[5] Furthermore, the scientific discoverability of the laws governing our universe is not something we need for our existence. Animals survive just fine without understanding the laws of nature. In other words, why are human beings able to discover the principles governing our universe, and why does what we think about the universe correspond to the way things actually are?

As stated by physicist Paul Davies,

> Most scientists just take it for granted that the world is both ordered and intelligible, and it's the intelligible part I find really quite extraordinary. Because it's one thing to accept that the universe is ordered, but ordered in a way that human beings are capable of understanding is an extraordinary thing. And so, the question naturally arises, what is the explanation for that?[6]

The fact that the laws of physics are intelligible and can be expressed mathematically indicates they were created by an intelligent Mind. This intelligibility and mathematical elegance in the laws of nature could have come only from a creative and omniscient Mind; in other words, God.

The founders of modern science (e.g., Isaac Newton) believed the universe was the product of a Mind, and was intelligible to beings like ourselves because the universe itself was the product of an intelligent Being.[7] For example, Nicolaus Copernicus wrote, regarding his attempt to understand the workings of our solar system, that he was motivated by a desire to comprehend what he called "the mechanism of the universe, wrought for us by a supremely good and orderly Creator . . . the system the best and most orderly artist of all framed for our sake."[8]

This leads to another key question: Why can the laws of nature be understood by humans and expressed mathematically? As physicist Robert Kaita said, "These laws that seem to be finite but describe a whole slew of phenomena has to be, at least for me, some compelling evidence for a Designer, a Mind that says I want to be known and I have created a universe that's understandable."[9] As stated in the Bible (the "Book of God's Words"), we can discover God by learning about our world (the "Book of God's Works"). Like the earliest scientists, many people today believe that science is uncovering and understanding the handiwork of God.

Even some scientists today recognize that the exquisite order and mathematical elegance of the physical laws governing the universe point to cosmic design. As physicist Paul Davies stated, "The temptation to believe that the universe is the product of some sort of design, a manifestation of subtle aesthetic and mathematical judgment, is overwhelming. The belief that there is 'something behind it all' is one that I personally share with, I suspect, a majority of physicists."[10]

And, Leonard Mlodinow, co-author of *The Grand Design* with Stephen Hawking, said, "Physics and mathematics can never explain where the laws and where logic came from. That's where we start from and if people want to say 'that must come from God,' we certainly don't argue against that."[11]

As we have seen, divine design is a most compelling explanation for the beginning of the universe, as well as for the harmony, elegance, and intelligibility of its governing laws of nature. We will next look at an even more remarkable feature of the laws of nature, namely, that they are "finely tuned" so that our existence is possible.

CHAPTER 3

THE UNIVERSAL FINE TUNER

Since the beginning of time, the universe has been governed by the laws of nature, many of which contain constants with fixed numerical values in their mathematical expression. Some examples of such "constants of nature" are the gravitational constant (G), which appears in Newton's law of gravity; the mass of the electron (m_e), and the "fine structure constant" (α), which determines the strength of electromagnetic forces. There are many other constants of nature. While these constants could have had any value whatsoever, one of the most striking and unexpected discoveries of modern science is that the laws and constants of nature are perfectly balanced "on a razor's edge" so as to make the universe habitable for life.[1]

For example, let's look more closely at the gravitational constant ("G" in Newton's formula $F = G\, m_1\, m_2\, /\, d^2$). It is equal to 6.67×10^{-11} N m^2 / kg^2.[2] Let's not worry about those metric units, but merely ask why it has that particular value when measured in those units. Why is the gravitational constant not 6.59 or 6.78, or some other value? Why is it 6.67 when measured in these units and how did it get that precise value?

And, let's also ask: What would happen if the gravitational constant had a slightly different value, while keeping all the other constants of nature as they are? It turns out that the universe would likely be so radically different that life would be impossible and we would not be here.

During the past forty years, scientists have looked at many of the constants of nature and asked what the universe would look like if they had different values. In many cases, they have found that their values are so sensitive if they were to change by just a tiny fraction, the universe as we know it would not be here and neither would we.[3] Since these constants are so critically balanced, they are often described as being "finely tuned."

Jay Richards, co-author of *The Privileged Planet* stated,

> If you were to take the basic fundamental constants of nature and you were to change these even slightly, or you were to pick their values at random, you would almost never get a universe that would be habitable in any sort of way. That is, you couldn't have galaxies, you couldn't have planets, and you could not have complex biological organisms if these fundamental constants were even slightly different, slightly stronger or slightly weaker, than they actually are in this universe. That's the idea of fine tuning.[4]

To better appreciate what this means, imagine a device with dials able to regulate the relative values of each of the fundamental forces of nature (i.e., gravity, electromagnetic forces, and strong nuclear and weak nuclear forces). If you change, even slightly, the relative strength of any one of these fundamental forces from its current setting, such as gravity, the impact on complex life would be catastrophic.[5]

For example, an alteration of the **gravitational force** by a mere one part in 10^{40} would mean that stars like our life-sustaining sun could not exist.[6] If the force of gravity was slightly stronger, only very large stars would form. However, since large stars "burn out" relatively quickly they have a life span too short for biological life to develop.[7] If gravity was slightly weaker, stars would either not exist or be too small to have enough energy to fuse hydrogen and form the heavier elements necessary for life, such as carbon.[8] Thus, the gravitational force constant must have the precise value that it actually does in order for stars like our sun to develop and have sufficient energy to support life.

This fine tuning is also found at the atomic level. Electrons are bound by the **electromagnetic force** into orbits around the atomic nucleus to form atoms, which are the building blocks of molecules. If the electromagnetic force had a different strength, this would change the number of elements that could exist in nature and, therefore, the kind of chemistry that could occur.[9] For example, if the electromagnetic force were a little stronger, heavier elements important for life, such as potassium, calcium, and iron, would not exist.[10] Thus, the complex biological molecules necessary for life to exist could not form and we would not be here.

The **strong nuclear force** binds protons and neutrons together in the nucleus of atoms. However, a very small

variance on either side of the actual value of the strong
nuclear force constant would have been disastrous for
the development of life. If the strong nuclear force had
been just a few percent higher, there would have been no
hydrogen in the universe after the Big Bang.[11] The resultant absence of heat from stars (fueled by the nuclear
fusion of hydrogen) and the lack of water (H_2O) in the
universe would have precluded the development of life.
On the other hand, if the strong nuclear force were just a
few percent weaker than it actually is, protons and neutrons would not "stick together" and there would be no
elements heavier than hydrogen (e.g., no carbon, which
is the building block of life).[12] Once again, life would not
have been possible.

And, if the **weak nuclear force** were a little smaller
than it actually is, then all the hydrogen in the early universe would have turned into helium.[13] This would mean
that there would be no water that is necessary for life and
no normal stars with a life-span long enough to support
life.[14] Conversely, if the weak nuclear force were a little
stronger, supernova explosions would not scatter the
heavy elements that are needed to form planets and are
necessary for life.[15]

More than just the fundamental forces of nature are
"finely tuned." Scientists estimate there are about twenty
physical constants and cosmological parameters that are
required to be calibrated precisely as they are in order to
have a universe capable of sustaining life.[16]

For example, the **cosmological constant** describes
the expansion speed of space in the universe. If space
expands too fast, the universe would spread out too

quickly for material objects to form. So, stars, galaxies, and planets would not exist. It turns out that if the rate of expansion of the universe right after the Big Bang had varied by as little as one part in a quintillion (a quintillion is a "1" with eighteen zeros after it), the universe would have continued to expand or collapse back on itself, and the world in which we live would not have been possible.[17] Laura Danly of Griffith Observatory stated,

> To illustrate just how small a number one part in a quintillion is, imagine all of the grains of sand in all of the world's beaches, that number is probably somewhere around a quintillion. In this analogy, if all of that sand represented the rate of expansion of the universe right after the Big Bang, how many grains of sand would I need to add or subtract to wreck the universe? Just one grain. One in a quintillion. That's how precise things had to be for us to be here.[18]

Brian Greene, author of *The Elegant Universe*, said of the many physical constants and cosmological parameters that are all calibrated "just right" for life to exist:

> These are numbers like the mass (the weight) of an electron, the weight of a quark, the strength of gravity, the strength of the electromagnetic force, about twenty numbers that describe those and other parameters, features of our world, but nobody knows why it is that those numbers have the particular values that they do. Now, you could easily say, "Who really cares? If you change the mass of

an electron by a little bit more or a little bit less, does it really matter?" And, the answer is: it does. It turns out if you imagine that we had twenty dials right here and we could fiddle with those twenty numbers at will, even a small change to the known values of those numbers would cause the world as we know it to disappear.[19]

Many other scientists have marveled at the finely tuned nature of our universe and its implications for our existence. For example:

- Oxford physicist Stephen Hawking has written, "The laws of science, as we know them at present, contain many fundamental numbers, like the size of the electric charge of the electron and the ratio of the masses of the proton and the electron. . . . The remarkable fact is that the values of these numbers seem to have been finely adjusted to make possible the development of life."[20]
- And, Nobel Prize-winning physicist Charles Townes has stated, "This is a very special universe: it's remarkable that it came out just this way. If the laws of physics weren't just the way they are, we couldn't be here at all."[21]

This "fine tuning" of the universe is also called the "Anthropic Principle" ("*anthropos*" means "human being" in Greek). The "Anthropic Principle" states that the seemingly arbitrary and unrelated constants in physics have one strange thing in common; namely, they all have precisely the values needed for the universe to be capable of producing and supporting life.[22] In other words, if any of these finely tuned "anthropic coincidences" were even slightly different, we would not exist.

Remember our imagined dials that control the twenty or so physical constants and cosmological parameters could have been set to any value whatsoever. However, all these "dials" are set just right in order for life to exist. If any one of them was even slightly different, we would not be here. Doesn't this seem a bit too unlikely to be just a random occurrence? Many think so.

A natural question many people ask when confronted with knowledge of this "fine tuning" is: Who designed it so that our "just so" universe fits together so perfectly to give rise to life? The conclusion that the universe had to be very precisely and finely tuned for life to exist seems suspiciously like a plan. Thus, a most compelling response is that the Creator of our universe set those numbers at just the right values so we could exist. The many "cosmic coincidences" leaves one with the overwhelming impression that the amazingly precise fine tuning of our universe was built in by a "Fine-Tuner"—God the Creator—who made the universe habitable for life.

How could such careful and precise fine tuning ever have been the result of mere chance? Physicist Mickey Kutzner doesn't think it was,

> Suppose we had twenty-five or thirty (stacked) walls that were standing in front of us and each wall was a mile or so long. Suppose you were told that each wall had a hairline crack in it, but that

it was totally random where that hairline crack would occur. Wouldn't it surprise you if you peered into one of those hairline cracks and you could see daylight through to the other side of those twenty-five walls? That would tell you that all of those hairline cracks were lined up perfectly. That's the same thing that we have in the universe with all of these physical constants adjusted just right for life to exist in the universe. The chances of any one of those are slim, the probability of all of them occurring at once, is extraordinarily unlikely.[23]

Dr. Francis Collins, Director of the National Institutes of Health (NIH) and former head of the Human Genome Project, states, "If one is willing to accept the argument that the Big Bang requires a Creator, then it is not a long leap to suggest that the Creator might have established the parameters (physical constants, physical laws, and so on) in order to accomplish a particular goal."[24] Such a goal includes making possible the development of intelligent life.

This evidence of a "just so" universe finely tuned for life is so powerful that it has led some scientists to question their atheism. For example, famous physicist Sir Fred Hoyle found these "anthropic coincidences" so striking that he abandoned his former atheism and declared,

A common sense interpretation of the facts suggests that a superintellect has monkeyed with physics, as well as with chemistry and biology, and that there are no blind forces worth speaking about in nature. The numbers one calculates from the facts

seem to me so overwhelming as to put this conclusion almost beyond question.[25]

And, Patrick Glynn wrote of scientists who believe
that life and the universe are accidental,

> Many scientists and intellectuals continue to cleave
> to this (atheistic) worldview. But they are increas-
> ingly pressed to almost absurd lengths to defend it.
> Today the concrete data point strongly in the direc-
> tion of the God hypothesis. It is the simplest and
> most obvious solution to the anthropic puzzle.[26]

Given the theological implications of our universe's
fine tuning, it is not surprising that some scientists are
questioning whether our universe is the only universe in
existence. Recently several scientists, most notably Stephen Hawking, have been promoting a theory of multiple universes, also called a "multiverse." This theory
attempts to explain the precisely fine-tuned nature of our
universe without resort to a God who "set all the dials"
just right for us to be here. These scientists propose that
our universe is only one of many universes, each with the
"dials" randomly set differently. They believe that we just
happen to be by chance in a universe with all the "dials"
set just right. In other words, we were just "lucky" and
won the "cosmic lottery."

However, a major shortcoming of this speculative
theory is that the presence of any other universes is
undetectable and unverifiable by the scientific method.
As physicist Paul Davies wrote,

> To invoke an infinity of other universes just to

explain one is surely carrying excess baggage to cosmic extremes. . . . It is hard to see how such a purely theoretical construct can ever be used as an explanation, in the scientific sense, of a feature of nature. Of course, one might find it easier to believe in an infinite array of universes than in an infinite Deity, but such a belief must rest on faith rather than observation.[27]

British astronomer and cosmologist Edward R. Harrison in his book, *Masks of the Universe*, concurred,

The fine tuning of the universe provides prima facie evidence of deistic design. Take your choice: blind chance that requires multitudes of universes, or design that requires only one. . . . Many scientists, when they admit their views, incline toward the teleological or design argument.[28]

In other words, it sure seems that our universe was purposefully designed by God so that life could develop and we could exist. Interestingly, it has also been discovered that the fine tuning of the universe extends to our solar system and planet Earth. This "local fine tuning" is what we will look at next.

CHAPTER 4

THE LOCAL FINE TUNER

Scientists have discovered many factors that are "just right" for advanced life on Earth. While there are many planets in the universe, the earth has an amazing array of factors that are aligned perfectly to be precisely what is needed to make complex life possible. Even though many of these factors may be present elsewhere, Earth seems to be a relatively special and uncommon planet where they are all aligned "just right" for advanced life.

A primary factor necessary for biological life is *liquid water* on or near the surface of the planet.[1] Water is called the universal solvent because it dissolves more compounds than any other liquid. Thus, water is an ideal medium for the chemistry of carbon-based life. Carbon has properties that make it uniquely suitable to build the complex molecules and structures necessary for biological life.

Of course, water in the liquid state is only possible if the planet is not too hot and not too cold. In other words, the planet needs to be at just the *right distance from its home star* if liquid water is to exist. For if a planet is too close to its star, water will boil away or vaporize. Too far and any water will be frozen.

The circular band around a star where liquid water is possible defines what scientists call the "circumstellar habitable zone." The earth just happens to be within this "Goldilocks Zone" around our sun where it is not too hot and not too cold for the presence of liquid water, which is necessary for biological life. Furthermore, for a planet to maintain water as a liquid, its orbit must be nearly circular so as to always remain within the habitable zone as it circles around its home star. Such is the case with Earth.

Within our solar system, the habitable zone is relatively narrow. It begins well outside the orbit of Venus and ends short of the orbit of Mars.[2] On Venus, our next closest planet toward the sun, the surface temperature is over 800 degrees Fahrenheit, which is a very hostile environment for life. Mars, our next closest planet away from the sun, is too cold for liquid water and complex life.[3] If Earth were just five percent closer to the sun, it would be subject to the same fate as Venus, with temperatures too high for liquid water and life to exist. Conversely, if Earth were twenty percent farther from the sun, carbon dioxide clouds would form in the upper atmosphere and surface temperatures would be too cold for complex life to exist.[4]

Another interesting characteristic of water is that it retains heat well. This peculiar property of water curtails extreme temperature swings in the earth's climate and moderates temperatures inside our bodies. And, even more oddly, unlike nearly all other compounds water is denser in the liquid state than in the solid state. This means that ice floats on top of liquid water instead of sinking to the bottom. This unique property of water prevents the earth's oceans, seas, lakes, and ponds from freezing solid, which would clearly be problematic for aquatic life.

Complex life also needs a **terrestrial planet**, like Earth, with a hard surface where liquid water can collect, not a gas planet like Jupiter or Saturn. Gas planets do not have a hard surface where liquid water can be deposited and living organisms can dwell.[5]

The presence of **heavy elements**, such as iron, nickel, copper, carbon, and oxygen, are also necessary for complex life. These heavy elements are forged inside the superhot and superdense cores of stars. Some of these stars explode as a massive supernova, which ejects these elements out into space. Gravity eventually pulls these elements together to form new stars and planets, as was the case with our solar system. But gas planets like Jupiter and Saturn do not have large concentrations of the heavy elements necessary for complex life. Terrestrial planets, however, do have large amounts of these heavy elements. Furthermore, for complex life to exist these heavy elements need to be present not only on the surface of a terrestrial planet, but also deep in its core, as is the case with Earth.

The movement of the liquid iron core deep inside the earth creates a **magnetic field** around our planet. This magnetic field is essential for the existence of complex life on Earth. The earth's magnetic field protects life from harmful radiation from the sun and prevents high-energy particles from outer space from stripping away the earth's atmosphere.[6] If our planet was smaller, its magnetic field would be weaker. A weaker magnetic field would allow the solar wind to strip away the earth's atmosphere and slowly transform our planet into a barren place like Mars.[7]

Another characteristic of a terrestrial planet like Earth is a hard, outer crust. Interestingly, scientists have found that the movement of the more than a dozen tectonic plates that comprise the relatively thin earth's crust is important for the existence of complex life on earth.[8] This movement is called **plate tectonics** and is needed to recycle the elements and minerals that life needs back to the planet's surface. Plate tectonics also helps regulate the temperature of the earth and creates the continents and shallow seas useful for life.[9] If our planet's crust was significantly thicker, plate tectonic recycling could not occur.[10]

Scientists have also found that the earth's **oxygen-rich atmosphere** contains about 21 percent oxygen, 78 percent nitrogen, and small amounts of carbon dioxide and other gases. This atmosphere is ideal for supporting complex life and enabling the mild climate necessary for liquid water.[11] And, the ozone layer in our upper atmosphere protects life on Earth by absorbing most of the ultraviolet radiation from the sun. Furthermore, the relatively low concentration of carbon in Earth's atmosphere prevents runaway greenhouse heating of the earth's surface, as is the case on Venus due to its carbon-rich atmosphere.

It is also interesting to note that scientists believe the high concentration of oxygen in our atmosphere is the result of the earliest life on earth. During the first 500 million years of its existence, the earth was too hot to be hospitable for life. However, microbial life was present a mere 150 million years after the earth cooled enough to allow life to survive.[12] Then, for over three billion years these single-cell organisms were the only form of life on earth. The amount of oxygen in the earth's atmosphere was minimal until about two billion years ago when the Great Oxidation Event (GOE) occurred.[13]

This relatively sudden jump in oxygen levels was due to single-cell organisms known as cyanobacteria, or blue-green algae. These microbes conduct photosynthesis, using sunshine, water, and carbon dioxide, to produce oxygen (as do plants today). Climate, volcanism, and plate tectonics played a role in regulating the earth's oxygen level during various time periods. But, one thing is clear—the origin of oxygen in the earth's atmosphere derives from life.[14] Scientists have also come to realize that oxygen levels needed to rise as high as they did to enable the development of more complex life forms, including animals and humans. Otherwise, we would not be here. So, we just happen to live at the right location in the solar system on the right type of planet, with the right magnetic field, and the right atmosphere for us to exist.

But there is still more. Earth's moon is about one-quarter the size of our planet. For a size of a planet like Earth, our moon is big. Scientists have come to realize that if the earth's *large moon* did not exist, neither would we.[15] That's because the moon's powerful gravitational pull stabilizes the tilt of the earth's axis of rotation at a nearly constant 23 degrees, relative to the plane at which the earth orbits the sun. This stable tilt of the earth's rotational axis stabilizes our planet's climate by enabling relatively mild seasonal changes and provides large areas of Earth's surface with a temperate climate that is hospitable to advanced life.

Without its moon, the earth would wobble on its axis. If our moon was not so large and so well-placed, the tilt of the earth's axis could vary by as much as 90 degrees during its annual revolution around the sun.[16] At a tilt of zero degrees, no sunlight would ever reach the poles. Whereas, at a tilt of 90 degrees, one pole would be subject to sunlight for half of the year while the other pole would be in darkness. Such fluctuations in the tilt of the earth's axis and resultant surface temperatures would make our planet quite inhospitable for complex life. Mars has two moons, but they are not large enough to stabilize the tilt of its axis, so it wobbles.[17] This also creates hostile, planet-wide dust storms that are not conducive to the development and survival of complex life.

Not only do we live at the right location in the solar system on the right type of planet, with the right magnetic field, and the right atmosphere, and at the right tilt due to having the right moon, but our *sun is the right type and size of star* for complex life to exist. The sun is not a typical star. Over 90 percent of stars are less massive than the sun.[18] If the sun was less massive, the circumstellar habitable zone would be smaller and the earth would have to be closer to the sun to remain within its boundaries and be warm enough to support advanced life.

However, if a planet gets too close to its home star, the increased gravity locks the planet's rotation into synchronization with its orbit. This means that there is one rotation around the planet's axis for every revolution around its home star. This causes one side of the planet to always face toward the star and the other side to always face away from the star.[19] This rotational lock causes one side of the planet to burn and the other side to freeze, resulting in surface temperatures not at all conducive to the existence of complex life.

Furthermore, as discussed in the last chapter, stars much larger than our sun have a life span too short for complex life to develop.[20] And, the vast majority of stars are binary stars or triplets. These are stars relatively close to each other and that orbit around each other. It is hard to imagine stable planetary orbits in such a system. In order to have stable planetary orbits, a solar system requires a single star that is not too variable, just like our sun.

Not only is the earth at just the right location relative to the right type of star, we are also **protected by the presence of giant planets** further out in our solar system. Recent computer simulations have determined that without the gravitational pull of Jupiter and Saturn the inner solar system would be visited more frequently by comets and wandering asteroids.[21] This computer modeling has revealed that these giant planets act as "an asteroid and comet catcher," preventing these harmful objects from impacting earth.[22] Devastating consequences to life occur when a large comet or asteroid hits the earth, as was the case with the extinction of the dinosaurs. Reducing that risk is important for the development and existence of complex life. As Jay Richards, co-author of *The Privileged Planet*, has stated,

Science has found that we have guardian sentinels
(Jupiter and Saturn) that protect us from life-ster-
ilizing bombardment by comets. These planets figure prominently in our existence.[23]

Not only are we on the right type of planet at the right
location in a solar system with the right type of star and the
right configuration of "guardian" planets, we are also in the
right kind of galaxy and at the right location in that galaxy to
enable complex life to exist. Galaxies occur in three shapes,
elliptical, irregular, and spiral. Most galaxies are elliptical.
However, elliptical galaxies contain older stars and have
low concentrations of the heavy metals needed for complex life.[24] Irregular galaxies are mostly located near larger
galaxies. This causes gravitational irregularities that do
not favor the development of life. However, our Milky Way
Galaxy is a spiral galaxy. Fortunately, spiral galaxies have
a lot of gas/dust clouds and young stars that provide the
right environment for life to occur.[25]

But even spiral galaxies have many dangerous locations, including the star-packed center of the galaxy with
black holes, supernovas, and stellar close encounters.[26]
Fortunately, we are located far from the center of our galaxy, in a relatively safe region of the galaxy between two
spiral arms.

In summary, scientists have discovered many factors
that are "just right" for the existence of advanced life on
earth, including being on a planet that:
- has liquid water on the surface,
- is within the circumstellar habitable zone of
  a planetary system around a single star,

- is a terrestrial planet with a molten core and solid crust, not a gas planet,
- has lots of heavy elements, such as iron and carbon,
- has a moving liquid iron core that generates a protective magnetic field,
- has a relatively thin crust that enables plate tectonic recycling of elements,
- has an oxygen-rich atmosphere that allows complex organisms to survive,
- has a large moon that can stabilize the tilt of its rotational axis,
- orbits the right type of star,
- is in a planetary system with giant planets that shield the inner planets from too many asteroid and comet impacts, and
- is at a safe location in the right kind of galaxy.

All of these factors must be present at one place and at one time in order for advanced life to exist. It has been estimated that the probability of all these factors being simultaneously present is about one in 10^{15} (a thousandth of a trillion).[27] To quote physicist Bijan Nemati,

> It's a number like that (one in 10^{15}) that you have to compare to the hundred billion stars that are in the galaxy. A hundred billion is a very large number, but a thousandth of a trillion is much, much smaller. On their face value, these probabilities are speaking. What they are telling us is this can't happen, or this is very unlikely to happen in the galaxy. And, that's where the evidence is pushing us.[28]

Let's consider how improbable is the occurrence of just one of these many factors, namely, the earth's relatively large moon. Again, scientists have found that our moon, which stabilizes the tilt of Earth's axis and moderates surface temperatures, is necessary for our survival. But, how did the moon form? Based upon the evidence, many scientists think that our moon resulted from a giant impact of a planet about the size of Mars with a primitive mostly-molten Earth over four billion years ago.[29] Most of this smaller planet was absorbed into the earth's core, but the considerable debris ejected by the impact were eventually consolidated by gravity to form our very large moon. If this favored theory is indeed correct, many aspects of this collision had to be just right in order to produce a moon large enough to stabilize the tilt of Earth's axis. This includes the composition, mass, and density of the colliding planet, as well as the exact point and angle of impact. It is also thought that this impact added iron from the colliding planet to the earth's core, resulting in the large liquid iron core that generates our planet's protective magnetic field. Obviously, it is quite remarkable that all this happened "just right" so that advanced life could exist on Earth.

* * *   * * *   * * *

However, none of this absolutely precludes the possibility of life existing elsewhere, especially primitive microbial life. And, some may claim that all these factors being "just right" for advanced life on earth could be merely a matter of probability and luck. However, it is much more difficult to explain in this manner why several of the factors necessary for advanced life on Earth also enable scientific discovery of the cosmos. In other words, the earth not only simultaneously satisfies the multiple conditions necessary for a habitable planet, but amazingly some of these same conditions also provide the best overall setting for making scientific discoveries.[30] In short, our planet is not only fine-tuned for life, it is also fine-tuned for discovery of the world beyond.

For example, the earth's oxygen-rich atmosphere is transparent, unlike the opaque carbon-rich atmosphere of Venus. In our solar system, only the earth's atmosphere can sustain advanced life and only the earth's atmosphere is transparent. We can see through it to the cosmos beyond. Especially since the invention of the telescope, Earth's transparent atmosphere has facilitated mankind's scientific discovery of the universe. This is just one example of the unexpected correlation between the conditions required for a habitable planet and the conditions required for scientific discovery.

As astrobiologist and co-author of *The Privileged Planet*, Guillermo Gonzalez noted:

> It is a remarkable coincidence that the kind of atmosphere that's needed for complex life like ourselves does not preclude that life from observing the distance universe. It's a surprise. It's something that you wouldn't expect just chance to produce. Why would the universe be such that those places that are most habitable also offer the best opportunity for scientific discovery? The discoverability

of the universe is something we didn't need for our existence. It's something additional to it. It seems then that whatever the source of the universe is, it intended that it contain observers who can discover.[31]

The interplay of several other factors necessary for life, namely the earth having a large moon at the right distance from the right size of home star, also happen to cause total solar eclipses. Total solar eclipses have permitted many scientific discoveries, including validating Einstein's General Theory of Relativity. Scientists confirmed Einstein's predictions by observing how the sun's gravity bent light from distant stars located behind the sun, which are only visible during a total solar eclipse. We can't help but wonder why, from our perspective on the surface of the earth, the size of the moon and the sun appear to be nearly identical. The sun is vastly larger than the moon, but the sun is at just the right distance away from us and the sun and our (relatively large) moon are just the right sizes relative to each other so that total solar eclipses are possible. Why is it that the only place in our solar system where a perfect solar eclipse can be observed is the only place where there are observers? Doesn't this seem to be more than a mere coincidence? Many think so.

As noted by Jay Richards,

> Yoking the conditions for life with the conditions for discovery is just the sort of thing that an intelligent agent would be interested in doing. If the universe was intended or designed to be discovered,

then this combination of habitable factors is just what one would expect. It would resolve our sense of surprise to realize that the universe is designed for discovery.[32]

It sure seems that we humans have been put on a planet at the correct location in our solar system, with a large moon, and with a transparent atmosphere so that we can discover the universe beyond. There are several other examples of this correlation between the factors necessary for intelligent life and the ability of that life to make scientific discoveries.[33] Why is that? Could it be, as we have been examining in this book, that by doing science we find compelling evidence of God's existence, creative power, and majesty? As stated by Lee Strobel, author of *The Case for a Creator,*

> I think God intentionally created a habitat for us that allows us to see Him through the creation that He has left behind. And, this habitat is conducive for us to do scientific research. It didn't have to be that way. But, it is. Why? Because I believe that by doing science, we find God.[34]

CONCLUSION TO PART I

In the first part of this book, we looked at scientific evidence of God's existence from cosmology, physics, astronomy, and planetary geology. More precisely, we initially discussed how the sudden and explosive origin of the universe at the Big Bang points to a Creator beyond space and time who brought the universe into being.

Then, from physics, we explored how the mathematically elegant and intelligible laws of nature point to the existence of a creative Mind behind the governing principles of the universe. And, we saw how these laws and the physical constants of our universe are "finely tuned" and need to be delicately balanced "just so" as they are in order for life to exist. For example, even an infinitesimally small change in the gravitational force would make the existence of stars like our sun, and hence complex life on earth, impossible.

Finally, from astronomy and planetary geology we examined the "local fine tuning" of the many factors that are "just right" for advanced life on earth, including the earth's life-sustaining magnetic field, oxygen-rich atmosphere, and large moon, and how many of these factors needed for advanced life also enable scientific discovery.

All of these recent scientific discoveries offer compelling evidence of the "fingerprint of God" in the cosmos, which as we will see in Part II of this book is evident in life, as well. However, no matter how compelling the scientific evidence pointing to God's existence, this evidence is not definitive proof of the existence of God. As physicist Mickey Kutzner stated,

> We can't use science to either prove or disprove the existence of God. What we can do is to look for evidence that suggests we are not here out of totally random events. And, to me, the arrangement, the special balancing of the fundamental constants, the fundamental forces of nature, tells me there is a Being there that has put this package together, which has constructed this universe. Someone outside the universe that was thinking about us and put it together for our benefit.[1]

Contrary to prevailing current opinion, there is no real conflict between science and faith, especially as proving or disproving the existence of a transcendent God is beyond the realm of science. As stated by astronomer Alex Filippenko of the University of California,

> Ultimately, the point is, if there is no way to scientifically test a hypothesis through experiments and observations, it's not truly a scientific hypothesis. And, since the question of the ultimate origin and the ultimate creator is fundamentally an un-testable question, it's really not part of science.[2]

As Fr. Robert Spitzer of the Magis Center of Reason and Faith points out in response to scientists who claim that science has shown that God does not exist,

> Science cannot disprove something which is beyond our universe and the reason is it's taking its data from within the universe itself. That is tantamount to saying, "Oh, the cartoon character is going to use data from the cartoon to disprove the cartoonist." It just simply can't be done.[3]

The *Catechism of the Catholic Church* (#159) reinforces this lack of conflict between science and faith when it states,

> There can never be any real discrepancy between faith and reason. Since the same God who reveals mysteries and infuses faith has bestowed the light of reason on the human mind, God cannot deny himself, nor can truth ever contradict truth. "Consequently, methodical research in all branches of knowledge, provided it is carried out in a truly scientific manner and does not override moral laws, can never conflict with the faith, because the things of the world and the things of faith derive from the same God."

PART I:
PERSONAL REFLECTION/
DISCUSSION QUESTIONS

1. If you or someone you know does not believe in God, what are some of the reasons why not?

2. If you believe in God, what are some reasons for your belief (other than that is how you were raised)?

3. How can having doubts about your faith be a good thing?

4. Which scientific evidence discussed in the first part
    of this book (i.e., the Big Bang, laws of nature, fine
    tuning of the universe, or "local" fine tuning of our
    planet Earth) did you find new or the most convincing
    evidence for the existence of God, and why?

5. How does this scientific evidence for the existence of
    God affect your belief in a creator God?

PART II

BIOLOGICAL EVIDENCE
OF GOD'S EXISTENCE

INTRODUCTION TO PART II

In this part, we examine evidence of God's existence from a biological viewpoint. We initially address whether or not evolution is compatible with faith in God, as this is a challenge to faith experienced by many—especially students. The sentiment of many evolutionary scientists was expressed by the famous paleontologist George Gaylord Simpson when he declared, "Man is the result of a purposeless and natural process that did not have him in mind. He was not planned."[1] If true, it would be hard to see how God had much to do with the evolution of life on earth and, therefore, how any evidence of God's existence can be found in biology.

Before discussing evolution, however, we need to understand whether or not the creation stories found in the Bible (especially Genesis 1:1–2:3) contradict with what science has discovered about the age of the universe and the development of life on earth. In other words, is Genesis meant to be taken literally as written or did the author(s) use another literary style to convey revealed truth?

After looking at Genesis, we examine the concept of evolutionary creation, which is also called theistic evolution. Christianity and many other faiths profess that God is actively involved in His creation. This belief is known as theism. Belief in an uninvolved, disinterested God who was only active in creation at the beginning is known as deism.[2] The interaction of God with the created order is called divine providence. In order to better understand evolution as God's providential method of creation, we will look at four types of causes: primary, intelligent, secondary, and natural. Many people see evolutionary creation as a reasonable synthesis between faith and science.

We then look at a conflicting view of evolution that is common within academia and the scientific establishment, which has been called evolutionary naturalism or atheistic evolution. Next, we explore how a layered explanation provides a more complete understanding of evolution, inclusive of God's role in the evolutionary process.

Lastly, we look at how the informational content of the DNA molecule is best explained by an intelligent cause and this provides biological evidence for the existence of God.

CHAPTER 5

THE TRUTH OF GENESIS

Before we discuss evolution, let's look at the creation stories found in the first book of the Bible—the Book of Genesis. The Church recognizes the important and unique nature of these creation stories. For example, the *Catechism of the Catholic Church* (#289) states, "Among all the Scriptural texts about creation, the first three chapters of Genesis occupy a unique place."

We need to properly understand, therefore, what the Genesis creation stories are trying to teach and how they are trying to teach it. As such, are the Genesis creation stories meant to be taken literally and as scientific and historical fact? If not, how are the teachings of Genesis to be understood?

To help answer these questions, the *Catechism of the Catholic Church* (#337, emphasis added) states, "Scripture presents the work of the Creator *symbolically* as a succession of six days of divine 'work,' concluded by the 'rest' of the seventh day." The Catechism also states that the biblical creation account uses "symbolic language" (#362) and "symbolism" (#375). It also says that "figurative language"

(#390) is used in the book of Genesis. This leads us to ask: What do the words "symbolically" and "figurative" mean in these quotes from the Catechism?

The Catechism answers this question earlier (#109-110), when it states,

> In Sacred Scripture, God speaks to man in a human way. To interpret Scripture correctly, the reader must be attentive to what the human authors truly wanted to affirm and to what God wanted to reveal to us by their words. In order to discover the sacred authors' intention, the reader must take into account the conditions of their time and culture, the literary genres in use at that time, and the modes of feeling, speaking, and narrating then current.

Thus, in the previous quotes from the *Catechism of the Catholic Church*, "symbolically" and "figurative" mean that the Genesis account of creation is primarily concerned with the meaning and purpose of God's creative work, and not specifically with the historical or scientific details of how it was accomplished. In other words, the *plot* of the Genesis creation stories need not be taken literally, but it is meant to be understood as conveying deeper religious truths.

Pope John Paul II made this clear in his General Audience on January 29, 1986, when he stated,

> Through the power of this word of the Creator's fiat, "let there be," the visible world gradually arises. In the beginning the earth is "without form and void."

Later, under the action of God's creative word, it becomes suitable for life and is filled with living beings, with plants and animals, in the midst of which God finally created man "in his own image" (Genesis 1:27). *Above all, this text has a religious and theological importance. It doesn't contain significant elements from the point of view of the natural sciences.*[1]

Similarly, Pope Benedict XVI stated in his General Audience on February 6, 2013 that, "The Bible isn't meant to be a manual of natural science."

This view of Genesis was reinforced by Cardinal Christoph Schönborn, General Editor of the *Catechism of the Catholic Church* (CCC), when he stated at a 2006 conference on creation and evolution with Pope Benedict XVI in attendance that,

The first page of the Bible is not a cosmological treatise about the development of the world in six solar days. The Bible does not teach us "how the heavens go, but how to go to heaven."[2]

Furthermore, this view of Genesis is not new. For example, St. Augustine of Hippo, who was one of the most influential theologians of the early Church, believed when reason (science) has established a finding about the physical world seemingly contrary statements in Scripture should be interpreted accordingly.[3] In his great work, *De Genesi Ad Litteram* (*On the Literal Interpretation of Genesis*, c. A.D. 415), St. Augustine spoke of the problems with taking a Biblical text literally if doing so contradicts what is known from reason and science (e.g., on the age of the universe and the development of life on earth),

> Usually, even a non-Christian knows something about the earth, the skies, and the other elements of this world, about the motion and orbit of the stars and even their size and relative positions, about the predictable eclipses of the sun and moon, the cycles of the years and the seasons, about the kinds of animals, shrubs, stones, and so forth, and this knowledge he holds to as being certain from reason and experience. Now, it is a disgraceful and dangerous thing for a non-believer to hear a Christian, presumably giving the meaning of Holy Scripture, talking nonsense on these topics; and we should take all means to prevent such an embarrassing situation, in which people show up vast ignorance in a Christian and laugh it to scorn. The shame is not so much that an ignorant individual is derided, but that people outside the household of the faith think our sacred writers held such opinions, and, to the great loss of those for whose salvation we toil, the writers of our Scripture are criticized and rejected as unlearned men.[4]

Nonetheless, some people believe the creation stories in Genesis are literally, scientifically, and historically accurate. They believe the universe was created only several thousand years ago, in six "days." However, as noted by St. Augustine, we can understand these stories in another way and still be faithful to the proper interpretation of the Bible in the Church.

The Genesis creation stories can be understood as a literary form similar to the parables of Jesus. What was the important truth conveyed in the parables of Jesus? Was it the plot and storyline? For example, was the "Prodigal Son" a real person and was Jesus describing actual events? Was there a real, live "Good Samaritan" who actually helped an injured Jew? Not necessarily. That's because the plot of a parable does not have to be literally and historically accurate in order for the parable to teach us something very important.

The truths that Jesus conveyed by means of parables are found within the plot. For example, God's love for each of us is the same as the unconditional love of the father towards his Prodigal Son. And, like the Good Samaritan, we are called to proactively demonstrate our love for all, even those with whom we differ or who are alienated from us. These are examples of profound religious truths being conveyed within the plot of the parables of Jesus.

Similarly, the Genesis creation stories are like a parable. The plot in the Genesis creation stories need not be literally and historically accurate in order for them to convey profound religious truths. Some of the religious truths in the Genesis stories of creation are:

1. God created the world from nothing (see CCC #296, 338).
2. God made the universe in an orderly manner over a period of time (see CCC #299, 341).

3. Everything God made is very good (see CCC #299, 302, 339). This is repeatedly emphasized in the Genesis creation stories.

4. Life appeared in ever-increasing complexity over a period of time (see Genesis 1:11, 20, 24, 26): First plants, then sea animals, and then land animals. Finally, human beings were created.

5. Each human being and all human beings together are the supreme culmination of God's creation and are created in God's own image and likeness (CCC #343, 396).

Thus, the biblical stories of creation are like a parable in that the plot does not have to be literally accurate (e.g., creation in six, twenty-four-hour "days") in order for them to convey profound religious truths (e.g., the orderly and increasingly complex nature of God's creative activity). But, while the Genesis creation stories are similar to a parable, there is a significant difference. A parable usually tells of fictitious events, whereas the Genesis creation stories describe real events in symbolic and figurative language. For instance, the world did have a beginning, the human race did have a beginning, God did (and does) confer spiritual souls on human beings, there was an original sin and fall from grace, etc.

Furthermore, although the ancient Hebrews were sometimes tempted to worship natural things as gods (idols), as did other nations and cultures that surrounded them, the Genesis creation stories affirmed that they did not have to either fear or worship nature. The Genesis creation stories reveal that nature is not a god, nor is any natural thing a god. Rather, the early Israelites were taught by Genesis that God is transcendent, over and apart from the natural world He created.

Moreover, none of these revealed religious truths conflict with what science has discovered about the origin and development of the universe and life on earth. Indeed, many of the religious truths in Genesis are supported by science. For example, the Big Bang supports the religious truth in Genesis that God created the universe from nothing. Furthermore, science has discovered that the universe developed in an orderly manner over a period of time. First were clouds of hydrogen that gravity pulled together to form stars, then planets formed from remnants of exploded stars, etc. Science also shows that life appeared in an ever-increasing complexity over a period of time. First plants, then sea animals, and then land animals came into existence, followed by human beings.

Despite these alignments between science and scripture, science cannot teach us about the goodness, the purpose, or the meaning of life. Only theology and philosophy can do that. The Bible, which deals primarily with religious truth, is necessary to complement and complete what science tells us about our origins. Science and religion complement one another and **both** are needed to fully understand the entire truth about creation. Religion helps us understand *why* things are the way they are. Science helps us understand *how* they got that way. Let's next see how this applies to the evolution of life on earth.

CHAPTER 6

THE EVOLUTIONARY CREATOR

We've seen how the Genesis creation stories need not be taken literally and that we need both science and religion to understand the full truth about creation. With that in mind, let's now look at evolution, starting with the concept of evolutionary creation. *Evolutionary creation, also called theistic evolution, is the idea that God ordained and sustained the gradual evolution of life on earth.*[1]

Evolutionary creation allows that traditional religious beliefs about God are compatible with the modern scientific understanding of biological evolution. Simply put, this position holds that there is a God, that God is the Creator of the universe and therefore all life within it, and that biological evolution is simply a natural process within that creation. In other words, it was God who established the framework and natural laws and processes that enabled the development of the vast diversity of life we see on earth today.

The idea of evolutionary creation is supported by Genesis 1:24, which states, "Let *the earth bring forth* all kinds of living creatures." Genesis does not say that God directly created plants and animals in their final form, only that they came forth from "the earth."

Pope Benedict XVI, in his Papal Address to the Pontifical Academy of Sciences in October 2008, affirmed the Church's understanding that evolution per se does not contradict belief in God's creative action when he stated: "Questions concerning the relationship between science's reading of the world and the reading offered by Christian Revelation naturally arise. My predecessors Pope Pius XII and Pope John Paul II noted that there is **no opposition between faith's understanding of creation and the evidence of the empirical sciences**."[2] (emphasis added)

As Pope John Paul II stated in his General Audience on July 10, 1985,

> The evolution of living beings, of which science seeks to determine the stages and to discern the mechanism, presents an internal finality which arouses admiration. This finality which directs beings in a direction for which they are not responsible or in charge, obliges one to suppose a Mind which is its inventor, its creator.[3]

And, as Cardinal Schönborn stated at the 2006 conference with Pope Benedict XVI, "The possibility that the Creator also makes use of the instrument of evolution is admissible for the Catholic faith."[4] Again, we need both science and religion to understand the full truth about the origin and development of life on earth. As such, let's briefly review what science says about the evolution of life on earth.

Charles Darwin in his classic book, *On the Origin of Species*, proposed that the diversity and complexity we now see in life on earth evolved over very long periods of time. Natural selection is the mechanism that Darwin proposed caused this evolution. Natural selection is the principle by which a slight variation of a trait, if useful, is preserved and passed on to succeeding generations. That's because an organism with a trait which provides it with a "functional advantage" (such as the size of finch beaks) in the struggle to survive will prosper and produce more offspring. In other words, individuals with greater fitness are more likely to survive and reproduce, while individuals with lesser fitness are more likely to die early or fail to reproduce. This optimization through variation and selection has been called "survival of the fittest."

Natural selection results in the characteristics that provide more fit organisms with a "functional advantage" replacing over time those characteristics of less fit organisms. This differential reproductive success leads to changes in the characteristics of a particular population of a species and over time populations that evolve to be sufficiently different eventually become a new species.

It is important to note that there are two distinct parts of Charles Darwin's theory of evolution; namely, (1) common ancestry and (2) the specific mechanism that drives the gradual evolution of life. The common ancestry of various forms of life, also called common descent, is the concept that different kinds of creatures share the same ancestral lineage. In other words, even though certain animals may be very different today, their hereditary lineage can often be traced back to a common ancestor.

For example, gerbils and giraffes are both thought to be descendants of a single type of animal from the far past. While this may be surprising, there is compelling scientific evidence in support of the common ancestry of life on earth.[5]

First, the fossil record shows that life on earth appeared in increasing complexity over long periods of time. The oldest rocks contain fossils of the only simplest forms of life, not the most complex. Only the newest rocks contain fossils of more sophisticated animals. There is very good correlation between the age of rocks and the types of organisms that are present.[6] The fossil record leads paleontologists to conclude that new life forms of increasing complexity gradually appeared throughout earth's history, and that these new life forms were formed by the modification of founding species.

The common ancestry of various forms of life has been further supported by recent advances in DNA genome sequencing. This is too complicated to get into detail here, but what is important to note is genetic evidence reveals many instances of different kinds of animals sharing a mutation or set of mutations in their DNA. Because of this, it can be concluded that this mutation occurred in a common ancestor and the descendants simply inherited it.[7]

The second part of Darwin's theory deals with the specific mechanism that he proposed has driven the evolution of life on earth. Darwin proposed that evolution is caused by natural selection acting on random variations. The later neo-Darwinian synthesis of Darwin's theory with genetics holds that these variations occur as a result of chance genetic mutations. This mechanism is real and explains many changes within a species, such as finch beak sizes or moth coloring. This type of evolution within a species is called micro-evolution.

Moreover, most scientists accept extrapolating this micro-evolutionary mechanism to fully explain macro-evolution, which is the development of entirely new species and life forms over very long periods of time. As we will see in the next chapter, this mechanism is typically seen as explaining the appearance of design in life without needing a designer, such as God. Thus, this second part of Darwin's theory can be perceived as problematic by some Christians.

In order to clarify how religion and science complement each other in understanding evolution, we need to realize the difference between a "primary cause" and a "secondary cause," as well as between an "intelligent cause" and a "natural cause." These various types of causes can be summarized as follows:

- A primary cause is the first cause, but not a sole cause of something else.
- A secondary cause is a dependent cause.
- An intelligent cause is due to the action of an intelligent agent (e.g., human or divine).
- A natural cause is either a random (chance) event or the necessary result of a governing law(s) of nature.

Think of it this way: When Michelangelo carved the *Pietà*, was it he or his chisel that did the carving? The answer is **both**. Michelangelo was the primary cause and the intelligent cause behind the design and creation of the statue. His tool, the chisel, was a secondary cause and a natural cause. It was a natural cause because the laws of nature govern what actions are needed to affect a certain shape in marble with a particular chisel.

Similarly, God is the primary and intelligent cause of the universe and all within it. But, in the ongoing evolution of the universe and of life, God prefers to work through secondary natural causes, such as natural selection acting on random genetic mutations.[8] God's creative actions are guided by the laws of nature He put in place. Therefore, evolution is merely God's "tool" (His "chisel," so to speak).

Charles Darwin acknowledged as much when he wrote, "To my mind it accords better with what we know of the laws impressed on matter by the Creator that the production and extinction of the past and present inhabitants of the world should have been due to secondary causes."[9] The *Catechism of the Catholic Church* (#308) similarly states, "God is the first cause who operates in and through secondary causes." Simply put, secondary and natural causes are also an expression of God's creative activity.

This concept was expressed in a 2004 report by the Church's International Theological Commission (ITC), while Cardinal Ratzinger (who later became Pope Benedict XVI) was its president. The ITC stated,

> God wills to activate and to sustain in act all those *secondary causes* whose activity contributes to the unfolding of the natural order which he intends to

produce. Through the activity of **natural causes**, God causes to arise those conditions required for the emergence and support of living organisms.[10]

Thus, evolution does not replace or negate God's creative activity in the origin and development of life on earth. Rather, *evolution is God's method of creation*. In other words, it was God's presence and guidance, subtly working through the natural laws and processes He created, that made possible the overwhelming complexity and diversity of life we see today. As stated by biologist Darrel Falk,

> If any lesson has come out of biology, it is that God works in subtle ways. God clearly uses natural forces to accomplish God's purposes, and often we see God's hand only when we look back in faith at the finished product. In doing so we worship God for a finished product that is awesome indeed.[11]

This view of the ongoing and developmental nature of God's creative activity is not new in the Church. For example, St. Augustine spoke of "primordial seeds, whence all flesh and all vegetation are brought forth" and eventually develop into the diversity of plants and animals we see today.[12] Augustine's reference to "seeds" is an analogy meant to compare the development of animal species to the bringing forth of trees, which start out as a seed and then grow into their mature form. St. Augustine wrote, "The things [that God] had potentially created . . . [came] forth in the course of time on different days according to their different kinds . . . [and]

the rest of the earth [was] filled with its various kinds of creatures, [which] produc[ed] their appropriate forms in due time."[13] This is evolutionary thinking. According to St. Augustine, God did not create all species of animals at once. He created them in a potential kind of way, so that they would develop over time, "producing their appropriate forms in due time." St. Augustine taught that animals, like trees, "did not spring forth suddenly in a mature size and form" but gradually "went through a process of growth" through which they "took their shape" and "developed all their parts."[14] St. Thomas Aquinas, for his part, repeatedly cited this view of St. Augustine's in his *Summa Theologica*.[15]

The *Catechism of the Catholic Church* (#302), too, implicitly supports evolutionary creation when it states,

> Creation . . . did not spring forth complete from the hands of the Creator. The universe was created "in a state of journeying" toward an ultimate perfection yet to be attained, to which God has destined it. We call "divine providence" the dispositions by which **God guides his creation** toward this perfection. (emphasis added)

For as Pope Benedict XVI proclaimed at his papal installation ceremony in 2005,

> We are not some casual and meaningless product of evolution. Each of us is the result of a thought of God. Each of us is willed, each of us is loved, each of us is necessary.[16]

Let's next look at the conflicting view of evolution referred to by Pope Benedict XVI, which holds that human beings are indeed a "casual and meaningless product of evolution." This commonly accepted view of evolution has been called evolutionary naturalism or atheistic evolution.

CHAPTER 7

NOT REVEALED BY EVOLUTIONARY NATURALISM

Many scientists are non-religious and so tend to interpret evolution through non-religious lenses. These scientists, therefore, do not accept the view of evolutionary creation discussed in the previous chapter. The general view of evolution within the scientific and academic communities is that life on earth developed without any involvement by God at all. For example, evolutionary biologist and Oxford professor, Richard Dawkins, author of *The God Delusion*, has written,

> Where does that [Darwinian evolution] leave God? The kindest thing to say is that it leaves him with nothing to do, and no achievements that might attract our praise, our worship or our fear. Evolution is God's redundancy notice, his pink slip. But we have to go further. A complex creative intelligence with nothing to do is not just redundant. . . . God is not dead. He was never alive in the first place.[1]

This understanding of evolution holds that life on earth originated and developed without any divine plan or guidance. This was the view of Francis Crick, co-discoverer of the double helix structure of the DNA molecule, who stated, "Biologists must constantly keep in mind that what they see was not designed, but rather evolved."[2] Richard Dawkins has also written that evolution makes it possible to be "an intellectually fulfilled atheist."[3] The widespread acceptance of this way of thinking is why evolution is often seen as a challenge to faith and why we are addressing it in this book.

But, what is going on here? Why does this view of evolution essentially hold that natural causes (e.g., natural selection acting on random genetic mutations) are the primary cause of the evolution of life, and not secondary causes as discussed earlier? Why is this atheistic view of evolution promoted by so many in the academic and scientific communities?

Cardinal Schönborn provides insight into what is happening when he wrote in *First Things*,

> We must reexamine the genuine science at work in Darwin's theory and its developments, and begin to separate it from ideological and worldview-oriented elements that are foreign to science . . . *modern evolutionary theory must be freed from its ideological shackles.*[4] (emphasis added)

What "ideological shackles" is he referring to? He's talking about a non-scientific, ideological, and philosophical underpinning of evolutionary science that is common today.

The general world view within the scientific and academic communities is that the material world of nature (matter and energy, space and time) encompasses the whole of reality. *This philosophical belief that nature is "all there is" is called materialism or naturalism.* This world view holds that the spiritual world does not exist or is unknowable.

This ideology was exemplified by astronomer Carl Sagan when he opened his famous 1980 TV series *Cosmos* with the statement, "The cosmos is all that is or was or ever will be."[5] Is this statement an established scientific fact? Or, is it merely his personal opinion? Yes, this statement is just a personal opinion based on his materialistic view of the world. That is because science, which deals only with observations in our physical and material world, can say nothing about the existence or non-existence of the spiritual world (heaven, hell, angels, our soul, etc.).

It is understandable for scientists to have a naturalistic perspective. Scientists are charged with pursuing a rational understanding of natural occurrences according to their best judgment. For a scientist to admit that "God did it" is seen as giving up on science and essentially "throwing up one's hands" rather than pursuing years of arduous study and research needed to try to fully understand the natural mystery being explored.

Moreover, many people view science as the only way to explain everything about our world and to bring God or faith into that discussion is not acceptable. Thus, it is not surprising that many in the scientific and academic communities embrace a materialistic ideology and naturalistic world view that seems to rule out the possibility that God had anything to do with the origin and development of life on earth.

The naturalistic view of the history of our universe essentially holds that *"Hydrogen is a light, odorless gas, which, given enough time, turns into people."*[6] Hydrogen was the primordial element after the Big Bang and the early universe was mostly hydrogen. Gravity then condensed that gas into stars, in whose cores hydrogen atoms were fused into heavier elements, including those necessary for life. Some of those stars exploded, ejecting the heavier elements out into space. Gravity then formed these elements into new stars and planets, including Earth. Then, primitive life appeared on our planet, followed by plant life, then sea animals, then land animals, and finally humans. So, in a sense, hydrogen did indeed turn into our human bodies.

Some scientists, however, would have us believe that hydrogen just turned into people without any divine involvement at all. But, is it reasonable to expect that hydrogen could have turned into people, even over very long periods of time, without any divine plan or guidance? *Surely, the chain of events that enabled hydrogen to turn into people had a primary and intelligent cause to ordain and sustain this most incredible hydrogen-to-humans process.*

The appearance of conflict between faith and evolution, therefore, arises not from Charles Darwin's theory of evolution per se, but from a misguided confusion of scientific theory with the atheistic world view of materialism and naturalism. *This confusing synthesis of evolutionary science and atheistic philosophy has been called "atheistic evolution," "evolutionism," or "evolutionary naturalism."*[7] And, it is this atheistic world view that is the "ideological shackles" from which evolutionary science needs to be freed.

Question #42 of the *Youth Catechism of the Catholic Church* (YOUCAT) states, "Natural science cannot dogmatically rule out the possibility that there are purposeful designs in creation."[8] YOUCAT then goes on to say, "A Christian can accept the theory of evolution as a helpful explanatory model, provided he does not fall into the heresy of evolutionism, which views man as the random product of biological processes."

Pope John Paul II also made this point in his General Audience on January 29, 1986, when he stated, "Indeed, the theory of natural evolution, understood in a sense that does not exclude divine causality, is not in principle opposed to the truth about the creation of the visible world, as presented in the Book of Genesis."[9]

As such, since we humans are confined by the "space-time continuum" it is impossible for us to preclude the possibility of any reality outside of that continuum. We're unlikely to be able to perceive the whole of reality because we are limited by our five senses and the tools we have created to enhance them.

For example, imagine if we did not have the sense of sight. How would we ever know the beauty of a magnificent, red sunset? That red sunset is reality. Nevertheless, how could we describe that reality, or the colors of a rainbow, to a person who has been blind since birth? For without our sense of sight we would be totally unaware of that reality. It is real, but we would not be able to perceive it. Similarly, it is entirely possible that if we had another sense (a sixth sense) we might be able to perceive more of reality than we can perceive with just our five senses.

By analogy, the fact that the physical world is accessible to us by the methods of science does not mean that the natural world is the whole of reality. We are too blinded by our human limitations to know for certain that the material world is the whole of reality.

Thus, any explanation of human origins that looks only to science is woefully inadequate. The origin of human beings, who have an integral unity of matter and spirit (body and soul), is not fully explainable by any purely scientific explanation.

Even though our material bodies may be the product of evolution, the Church holds that our immaterial and immortal soul, that which makes us truly human, is directly imparted by God. Pope John Paul II stated in his 1996 address to the Pontifical Academy of Sciences, "If the human body takes its origin from pre-existent living matter, the spiritual soul is immediately created by God."[10] The *Catechism of the Catholic Church* (#366) also states that "every spiritual soul is created immediately by God."

This was also clarified by Pope John Paul II in his General Audience on April 16, 1986, when he stated,

> From the viewpoint of the doctrine of the faith, there are no difficulties in explaining the origin of man in regard to the body, by means of the theory of evolution. . . . It is possible that the human

body, following the order impressed by the Creator on the energies of life, could have been gradually prepared in the forms of antecedent living beings. However, the human soul, on which man's humanity definitively depends, cannot emerge from matter, since the soul is of a spiritual nature.[11]

Therefore, the first true humans (body and soul) came into being when God imparted souls into their pre-existing bodies. Michelangelo beautifully portrayed this moment of creation on the ceiling of the Sistine Chapel when he showed God's finger reaching out and touching Adam's finger, thereby passing on the "divine spark of life" (our soul) that makes us human beings into the image and likeness of God.

In short, *the core distinction between evolutionary creation (theistic evolution) and evolutionary naturalism (atheistic evolution) is the philosophical conviction as to whether or not God is behind the evolutionary process*. Furthermore, the Catholic Church holds that even natural processes based on chance (contingency), such as random genetic mutations, are fully consistent with God's guidance and divine creation—even if God's role is not detectable by science.

In affirming that God can providentially work through what seem to be "chance" or "random" events, the 2004 report of the International Theological Commission stated,

> According to the Catholic understanding of divine causality, true contingency in the created order is not incompatible with a purposeful divine

providence. . . . Thus, even the outcome of a truly
contingent natural process can nonetheless fall
within God's providential plan for creation. . . .
In the Catholic perspective, neo-Darwinians who
adduce random genetic variation and natural
selection as evidence that the process of evolution
is absolutely unguided are straying beyond what
can be demonstrated by science. Divine causality
can be active in a process that is both contingent
and guided. Any evolutionary mechanism that is
contingent can only be contingent because God
made it so. An unguided evolutionary process—
one that falls outside the bounds of divine providence—simply cannot exist.[12]

Francis Collins, director of the National Institutes of
Health (NIH) and former head of the Human Genome
Project, addresses God's role in even chance events during
the evolution of life, such as random genetic mutations,
when he writes in his book *The Language of God*,

If God is outside of nature, then He is outside of
space and time. In that context, God could in the
moment of creation of the universe also know
every detail of the future. . . . In that context, evolution could appear to us to be driven by chance,
but from God's perspective the outcome would be
entirely specified. Thus, God could be completely
and intimately involved in the creation of all species, while from our perspective, limited as it is by
the tyranny of linear time, this would appear a random and undirected process.[13]

Once again, it is important to emphasize that evolution per se is not the main challenge to faith in God. Rather, the main challenge to faith is how evolution is often taught in schools and colleges, as well as how it is usually presented in the media. That is because evolutionary naturalism (atheistic evolution) is the basis of what is typically taught and presented on television. However, most scientists, teachers, and television programs on evolution don't call it "atheistic" evolution, even though that is what they mean. Rather, they just say "evolution" as if there is only one type of evolution, ignoring theistic evolution or evolutionary creation.

Thus, Christians need to be alert for and critical of the materialistic philosophy and naturalistic world view that underlies how evolution is often presented in schools, colleges, and on television. This atheistic philosophy refuses to acknowledge the possibility of God's creative action in the evolution of life. As Pope John Paul II stated at his General Audience on March 5, 1986,

> It is clear that the truth of faith about creation is radically opposed to the theories of materialistic philosophy. These view the cosmos as the result of an evolution of matter reducible to pure chance and necessity.[14]

As the *Catechism of the Catholic Church* (#295) states,

> We believe that God created the world according to his wisdom. It is not the product of any necessity whatever, nor of blind fate or chance.

And, as Cardinal Schönborn wrote in an op-ed letter printed in the *New York Times*, "Any system of thought that denies or seeks to explain away the overwhelming evidence for design in biology is ideology, not science."[15] Let's next look at a more complete and non-ideological way to understand evolution, including God's role in the evolutionary process.

CHAPTER 8

FOUND IN A LAYERED EXPLANATION OF EVOLUTION

A more complete way to understand evolution is found in what is called a "layered explanation."[1] A "layered explanation" means that things can be explained by multiple levels of understanding, in a plurality of compatible explanations.

This can be understood by considering the question, "Why is the pot of water boiling?" At a physical level, the boiling water in a pot can be explained by the increasingly rapid movement and collisions of water molecules. At another level, it can be explained by the fact that someone turned on the stove. And, at a still deeper level the boiling water can be explained by someone's desire for a cup of tea.

It is important to note that these various levels of understanding need not compete with each other. It is not a matter of being forced to accept one rather than another explanation. All of these explanations are true and they can happily coexist without conflicting with each other.

Similarly, the evolution of life can also be understood by a plurality of compatible layers of explanation. Let's consider the question: "Why are there so many species of life?" One answer could be because of dramatic events in natural history (e.g., climate change or an asteroid colliding with earth). And, on a Darwinian level there are so many species because of natural selection acting on random genetic mutations. Furthermore, at a deeper theological level there are so many species because of divine creativity, wisdom, and causation.

The deeper "I want a cup of tea" level could be considered equivalent to the role of God's divine providence and wisdom in ordaining and sustaining the evolution of life on earth; whereas, the movement of water molecules in the kettle can be seen as equivalent to a purely materialistic understanding of the evolution of life.

Again, these various levels of understanding need not compete with each other. The theological explanation need not conflict with the natural history or Darwinian explanations. All of these explanations of life's diversity are valid and they each can coexist without conflicting with one another.

Furthermore, it is important to note that the most meticulous scientific examination of molecular movement in the pot will never reveal the "I want a cup of tea" reason for the boiling water. Similarly, a purely naturalistic and materialistic understanding of evolution will not reveal God's role in the origin and evolution of life. Even the most exhaustive scientific examination of natural processes cannot prove the absence of divine causality.

In another analogy, consider Shakespeare's play *Romeo and Juliet* in which the tragic deaths of these two young lovers brings a reconciliation between their feuding families. Why did these two families stop feuding? Was it because of the deaths of Romeo and Juliet, or because Shakespeare wrote the play that way? **Both** are causes, but on different levels. The deaths of Romeo and Juliet are causes **within** the plot of the play. However, Shakespeare was the primary intelligent cause of these events in that he conceived everything in the play, including the characters and all their actions in the play's plot.

So it is with God's creative activities and the natural workings of our universe. Just as the author of a play is not visibly seen within the plot of a play, so too the "Author" of the universe is not visibly seen within the natural workings of the universe. But, that doesn't make the author's actions any less real. Rather, the events in the play's plot time and the author's writing activities, which are outside the play's plot time, are on two different levels of causality. Therefore, science, which operates in the natural world's "plot time," cannot definitively rule out that God could be the primary and intelligent "Author" of the universe and life on earth.

Similarly, we cannot understand a painting by only examining the brush strokes made by the artist on the canvas. If we did just that, we would totally miss the "big picture" of what the artist was trying to show and to what purpose. Likewise, evolutionary naturalism seeks to understand the details of **how** life evolved (the "brush strokes"). But, evolutionary naturalism cannot address the "big picture" of *who* caused that evolution to occur (the divine "Artist") or *why* we are here.

Therefore, evolutionary theory, which is based on the methods of natural science, cannot exclude the possibility that a divine "Artist" is the primary and intelligent cause of the origin and development of life on earth. So, let's next see if there is any biological evidence of an intelligent cause for the origin and evolution of life. After all, science identifies intelligent causes in other areas; for example, archaeology, forensics, and the Search for Extraterrestrial Intelligence (SETI). Why not also in biology if that is where the evidence leads?

CHAPTER 9

THE CAUSE OF GENETIC INFORMATION

Thus far in Part II, we have discussed how the Genesis creation stories teach religious, not scientific, truths and how evolution does not disprove the existence of God, contrary to the claims of some evolutionists. Rather, there are valid reasons to accept that the evolution of life on earth was God's method of creation. In that regard, we discussed how God is the primary intelligent cause of the universe and all within it. But, in the ongoing evolution of the universe and of life, God prefers to work through secondary natural causes.

Let's now look more closely at biology and see if there is scientific evidence of an *intelligent cause* for life on earth. We will do this by looking at what some scientists consider to be evidence of God's existence found within virtually every cell of all living things; namely, that life is built upon biological information.

Cells are miniature factories composed of submicroscopic molecular machines, which in turn are mostly composed of proteins. Proteins are worker molecules that

perform a vast array of key biological functions and partic-
ipate in virtually every process within cells. Many proteins
are enzymes that catalyze metabolic reactions. Proteins
also have structural or mechanical functions, such as
forming the "scaffolding" that maintains the cell's shape.

Proteins are composed of long chains of smaller
organic compounds called amino acids. Thus, amino
acids are one of the basic building blocks of life. However, *amino acids form useful proteins only when joined
together in the proper order.* It is important to note that
amino acids cannot self-assemble themselves into the
proper order required to make proteins.

Rather, biologists have discovered that deoxyribonu-
cleic acid (DNA) instructs the process of protein formation in cells. DNA contains the genetic instructions used
in the formation of proteins from individual amino acids.
Along with ribonucleic acid (RNA) and proteins, DNA
is one of the three major macromolecules in the cell that
are essential for life.

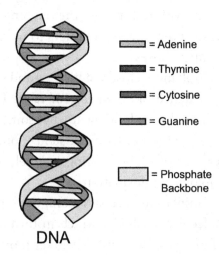

DNA

FIGURE 1

As illustrated in Figure 1, DNA consists of a double helix backbone (spine) made of sugars and phosphates. Attached to each strand of the backbone is one of four types of molecules called nucleotide bases, or simply "bases." The four nucleotide bases in DNA are *adenine (A)*, *cytosine (C)*, *guanine (G)*, and *thymine (T)*. The details of all this are beyond the scope of this introductory book, but for our purposes it is important to note that these nucleotide bases can only join together in matching pairs of adenine (A) with thymine (T) and guanine (G) with cytosine (C), as shown in Figure 1.

The specific sequence of nucleotide bases, arranged along the spine of DNA's helical strands, determines the specific order of amino acids to be assembled to make a protein. This sequence is critical. That's because the function of a protein depends entirely upon the specific order of the many amino acids of which it is composed.

We will next briefly explore how proteins are formed in the cell from amino acids.[1] Before doing so, please refer to Figure 2 to become familiar with some key terms in the cellular process by which DNA determines the specific amino acids that will be joined together in the proper order to make proteins. It is important to emphasize that nucleotide bases can only join together in matching pairs of *G with C* and *A with T*, as shown in Figure 2, and that they do so in groups of three nucleotide bases (triplets).

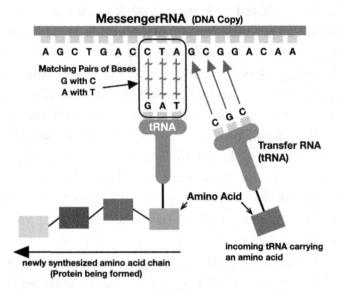

FIGURE 2

[Note: In RNA, thymine (T) is actually replaced with Uracil (U), which pairs in the same way as thymine (T). Therefore, A actually pairs with U (not T) in RNA. But, for simplicity let's assume T is in both RNA and DNA.]

The process of protein formation within a cell involves the addition of one amino acid at a time to the end of a newly forming protein chain, as shown at the lower left in Figure 2. As noted earlier, functional proteins are made only when amino acids are joined together in the proper order. And, it is the sequential arrangement of nucleotide bases in messenger RNA, which have been copied from a section of the cell's DNA, that determines which type of amino acid is placed next in the chain of amino acids being formed into a new protein. The amino acid to be added to the protein chain is determined by the next three-nucleotide sequence (triplet) in the messenger

RNA. In Figure 2, this next triplet in the messenger RNA at the top is shown as GCG.

Transfer RNA (tRNA) molecules convey the amino acids to be assembled into proteins. Only one type of amino acid can be attached to each type of transfer RNA molecule. At one end of the transfer RNA is the three-nucleotide sequence (triplet) corresponding to the specific type of amino acid that is attached at the other end. In Figure 2, the triplet CGC is shown at the top of the next (rightmost) transfer RNA molecule, with its amino acid attached at the other (lower) end.

Each amino acid is added to the protein chain by the transfer RNA with the complementary three-nucleotide sequence that matches the next triplet of the messenger RNA. For example, G+C, C+G and G+C as shown in Figure 2 (remember that G only pairs with C). In this way, the sequence of nucleotide bases in the messenger RNA determines the sequence of amino acids in the protein being formed.

Again, it is important to note that the next three nucleotide bases (triplet) in the messenger RNA molecule must correctly match the three nucleotide bases (triplet) in the next transfer RNA molecule (e.g., G+C, C+G, G+C, as shown in Figure 2). Only then will the correct amino acid join the chain of amino acids being formed into a new protein.

This is a much more complicated process than what we have covered here. But, for our purposes let's ask a very important and basic question. Namely, *how did the nucleotide bases in DNA (which are copied into messenger RNA) get into the correct sequences needed to join amino acids together in the proper order to make proteins?*

In other words, how did the As, Ts, Cs, and Gs in DNA (which is copied into messenger RNA) get into the proper sequence to correctly match with the corresponding nucleotide bases in the next needed transfer RNA with its attached amino acid? For example, if the next three nucleotide bases (triplet) in the messenger RNA molecule shown at the top of Figure 2 were G<u>A</u>G instead of G<u>C</u>G (as shown), the next needed amino acid, which is attached to a transfer RNA with its corresponding CGC triplet, would *not* be joined to the forming protein in the proper order, since A will not pair with G. In this case, the protein molecule that would be formed will likely be defective and not function properly.

Basically, the nucleotide bases in the DNA molecule convey instructions for the proper formation of proteins because of their specific sequence. Similarly, the letters in the human alphabet convey a meaningful sentence only when arranged in a specific sequence. *Just as the meaning of an English sentence depends upon the sequential arrangement of individual letters, so too the function of a protein depends upon the sequential arrangement of individual amino acids.*

Amino acids alone do not make proteins, any more than letters alone make words, sentences, or poetry. In both cases, the specific sequence of the constituent parts determines the function.[2]

The specific arrangement of letters in a meaningful sentence contains and conveys human encoded information. So, too, the specific arrangement of nucleotide bases in DNA contains and conveys biologically encoded information, namely, the precise set of instructions for building proteins. In other words, the sequencing of amino acids to make proteins is directed by the information—the set of biochemical instructions—encoded in the DNA molecule.[3] Without that information, life could not exist.

The specific arrangement of nucleotide bases along the spine of the DNA molecule stores and processes the protein assembly instructions—the information—in a four-character digital code (A, T, C, G). This is analogous to a computer that stores and processes information based on the specific arrangement of just a two-character digital code (0 and 1). In fact, Microsoft founder Bill Gates has written, "DNA is like a computer program, but far, far more advanced than any software we've ever created."[4] Evolutionary biologist Richard Dawkins has also acknowledged, "The machine code of the genes is uncannily computer-like."[5] That's why the DNA molecule is routinely described as an "information-rich system."

Likewise, James Watson, co-discoverer of the genetic code with Francis Crick, was asked to summarize the significance of their discovery in a single sentence. Watson thought hard and then replied, "All life is digital information."[6]

Furthermore, the specific sequence of nucleotide bases in DNA is not generated by chemistry or by any physical law. Because chemical bonds do not determine the sequential arrangement of nucleotide bases in DNA, these "letters" can assume a vast array of possible sequences and thereby express many different biochemical messages.[7] And, just like the information contained in a meaningful sentence does not derive from the chemistry of paper and ink, so too the information in DNA transcends the properties of its physical medium. Indeed, printed information originates from a source beyond the paper and ink that is used to contain and convey the information. The same is true of the information in DNA because the medium of chemistry and physics is not sufficient to determine the message (genetic information). Again, *the message transcends the medium and chemistry does not determine the code.*

Random genetic mutations (caused by DNA copying mistakes) can and do change the information in pre-existing DNA. And, even if just one nucleotide base is changed in the coding sequence of DNA, the result can be significant. A striking example of this is sickle-cell disease. This genetic disorder was caused by the mutation of a single nucleotide, from GAG to GTG.[8] Just replacing one "A" with a "T" in the coding sequence causes a change in the protein and makes red blood cells form a "sickle" shape. Biologists refer to such a change in only one nucleotide base as a "point mutation."

Of course, point mutations within an organism can be productive, resulting in a gain of function. Indeed, this is the fundamental basis of the natural selection process, where genetic changes (e.g., point mutations) that yield a functional advantage are preserved. Biologists typically extend this argument to suggest that numerous functional mutations may even explain the formation of the first cellular life. However, to create a totally new set of information is a very different thing than to slightly modify existing information. In other words, it is not so difficult to see how a random change in an existing life form may, on occasion, be beneficial. It is another thing entirely to suggest that randomness could somehow create information where there had been none previously.

Furthermore, just because point mutations look to us like totally random events, does this preclude any role for God in this process? Or, is it possible that what appears to us to be a random change could have been subtly guided by God to achieve the creative action in development of life that He desires? In other words, could God be intimately involved in the evolutionary creation of life by "tweaking" these point mutations in a way that is not detectable by science? How can this possibility be ruled out when our perspective is limited by the "tyranny of linear time"?

Moreover, an *intelligent cause* is known to be capable of generating useful information, such as a meaningful sentence. As the pioneering information theorist Henry Quastler observed, "Information habitually arises from conscious activity."[9] Thus, *the discovery of biological information in DNA provides strong grounds for inferring that intelligence had a causal role in the origin of this information.*

An intelligent cause is, therefore, a valid explanation for the biological information encoded in DNA. In other words, information-rich biological systems, such as DNA, provide compelling evidence for an intelligent cause of life. The specifically arranged and functionally significant sequences of nucleotide bases in DNA—the encoded information—implies the past action of an intelligent cause, even if such intelligence cannot be directly observed today.[10]

Former atheist Anthony Flew came to accept that the origin of the language-like "coded chemistry" of DNA cannot be explained without reference to an intelligent cause.[11] That is because meaningful information just does not spontaneously emerge from collections of mindless molecules.

In his book, *Signature in the Cell*, Stephen Meyer writes,

> Indeed, our uniform experience affirms that specified information—whether inscribed hieroglyphics, written in a book, encoded in a radio signal, or produced in a simulation experiment—always arises from an intelligent source, from a mind and not a strictly material process. So the discovery of the specified digital information in the DNA molecule provides strong grounds for inferring that intelligence played a role in the origin of DNA.[12]

In short, it is not a case of life evolving from either natural causes OR intelligent causes. Why could it not have been both?

Rather, the fundamental question today is whether natural causes (such as natural selection acting on random genetic mutations) are primary causes in and of themselves alone. Or, are natural causes merely secondary causes ordained and sustained by an intelligent agent, which is the primary cause?

We know that Michelangelo was the primary and intelligent cause who guided his chisel, which was merely a secondary and natural cause of the Pietà. In the same way, the information in DNA provides biological evidence that God was the primary and intelligent cause who guided the mechanism of natural selection acting on random genetic mutations, which was merely a secondary and natural cause of the evolution of life on earth.

The biological evidence of God's existence discussed in this chapter can be summarized in what could be called the Biological Information Argument, as follows:

1. Premise One–DNA contains useful information (to place amino acids in the proper order to make functional proteins).
2. Premise Two–Useful information habitually arises from an intelligent cause.
3. Conclusion–Therefore, an intelligent cause is a valid explanation for the useful information contained in DNA.

Indeed, Francis Collins, former director of the Human Genome Project, has called the genetic code the "language of God," without which evolution would not have been possible. In his book, *The Language of God*, he marvels that the human genome was "written in the DNA language by which God spoke life into being. I felt an overwhelming sense of awe in surveying this most significant of all biological texts."[13]

CONCLUSION TO PART II

In this part, we first looked at what the Genesis stories of creation say and how they say it. We saw how the Genesis creation stories need not be taken literally and are not meant to convey scientific truth regarding how creation occurred. Rather, the Genesis creation stories use a literary style similar to a parable to convey deep religious truth behind the symbolic plot line of the story. And, we discussed how we need *both* science and religion to understand the full truth, since science helps us understand the *how* and religion the *why*.

We then discussed the concept of evolutionary creation or theistic evolution, which is a reasonable balance between faith and science. This included looking at four types of causes to better understand how evolution was God's method of creation. In short, God is the primary intelligent cause of the universe and all within it. However, God prefers to work through secondary natural causes in the ongoing evolution of the universe and of life on earth.

Next, we saw that the general view of evolution within academia and the scientific establishment is based

111

on a materialistic philosophy and naturalistic world view. This limited understanding of evolution has been called evolutionary naturalism or atheistic evolution, and is often seen as a challenge to faith in God as it seems to rule out the possibility that God had anything to do with the evolution of life on earth.

We next discussed how a layered explanation provides a more complete understanding of evolution, including God's role in these processes. Finally, we examined how the genetic information encoded in DNA is explained by an intelligent cause and this provides biological evidence of God's existence.

In closing, let's hear again from Pope Benedict XVI who wrote (before he became Pope) in his book, *In the Beginning*:

> It is the affair of the natural sciences to explain how the tree of life in particular continues to grow and how new branches shoot out from it. This is not a matter for faith. But we must have the audacity to say that the great projects of the living creation are not the products of chance and error. . . . The great projects of the living creation point to a creating Reason and show us a creating Intelligence, and they do so more luminously and radiantly today than ever before. Thus we can say today with a new certitude and joyousness that the human being is indeed a divine project, which only the creating Intelligence was strong and great and audacious enough to conceive of. Human beings are not a mistake but something willed; they are the fruit of love.[1]

PART II:
PERSONAL REFLECTION/ DISCUSSION QUESTIONS

1. Do you believe the Genesis stories of creation are literally true and that God created the universe in six (24-hour long) days? Why or why not?

2. What scientific and/or religious truths do you believe the Genesis stories of creation are meant to convey?

3. Do you think evolution is compatible with faith in God? Why or why not?

4. Do you think evolution was God's method of creation? Why or why not?

5. What will you be alert for when learning about evolution (e.g., in school or on TV)?

6. Do you believe the genetic information found in DNA provides biological evidence of God's existence? Why or why not?

HUMAN EVIDENCE OF GOD'S EXISTENCE

INTRODUCTION TO PART III

In the first part of this book, we reviewed scientific evidence of God's existence from cosmology and physics, specifically the Big Bang, the laws of physics, and the "fine tuning" of the universe for life. In Part II, we looked at evidence of God's existence from biology, including how evolution is indeed compatible with faith in God. In this part, we leave science behind and examine evidence of God's existence from within human nature and by human reason.

To introduce this topic, consider this question: "If God created such a wondrous and perfect universe (as discussed in Part I), why is human society so flawed?" This question leads to an even-deeper form of evidence of God's existence. However, this evidence will not come from science, but from within ourselves, including our thoughts and feelings.

CHAPTER 10

THE CORE OF CONSCIOUSNESS

The enhanced consciousness found in human beings, namely our self-awareness and reasoning ability, is not necessary for survival. Mere instinct, which guides behavior in animals, is sufficient. Thus, there does not seem to be a need for living creatures to have attained the self-awareness and reasoning ability found in human beings in order to survive.

Rather, the nature of our enhanced consciousness suggests that humans are destined for more than mere survival. Our survival and physical flourishing are not enough. We want more. For example, only humans seek to discover the scientific principles that govern the universe. We seek to understand the underpinnings of our world and how things work. But, do any of us ever feel like we fully understand everything about everything such that we have no desire to learn more? Nope. We always want to learn more.

And, have you, your family, or friends ever felt frustrated in life without understanding why? Have you or others achieved success, but still feel unfulfilled? Though we may have achieved "success" by the standards of this world, do we still feel an inner emptiness and does this cause us to continue to strive for something more? Why?

The Five Transcendentals

The presence of our enhanced consciousness not only differentiates humans from animals, it also aids in making the case for the existence of God. In part, that is because our enhanced consciousness causes us to desire and yearn for five ultimate transcendentals that go beyond our need for survival. These five transcendentals are our desire for: (1) perfect knowledge/truth, (2) perfect love, (3) perfect justice/goodness, (4) perfect beauty, and (5) perfect home/being. Most interestingly, any temporal satisfaction of these inner desires or yearnings results in us feeling frustrated and still wanting more. That's because we seek a perfect experience of each of these five transcendentals.

For example, human beings do not seek just practical knowledge (e.g., "How do I get the food I need to survive?"). Rather, we want to know just for the sake of knowing, and **we have an innate desire for a complete and perfect explanation**. This is evident in the ongoing work of science in seeking a more and more complete understanding of our world. We know we have not yet reached a perfect and complete understanding of our world. Moreover, this lack of perfect knowledge leaves us feeling unfulfilled, wanting more. For this reason, we continue to seek even more knowledge, more truth.

Philosophers, including Plato and St. Augustine, have long wondered how we can desire **perfect** knowledge, love, justice, beauty, and home without being aware of them. They do not believe that the perfect can be abstracted from the imperfect. As such, our awareness of these perfect things cannot come from our imperfect world. Indeed, they believed the opposite. For example, without an awareness of perfect beauty we would not be able to see any imperfection in beauty. The same would be true if we did not have an awareness of what constitutes perfect love, knowledge, etc. Our awareness of what is perfect seems to require something perfect to **incite** it.

Priest, philosopher, and theologian Fr. Robert J. Spitzer, S.J., Ph.D. has written several books (especially *New Proofs for the Existence of God*) about how our conscious—but ultimately unfulfilled—yearning for more, especially these five transcendental qualities, provides evidence within human nature of the existence of God. Of all living creatures only human beings have an innate and conscious knowledge of and desire for perfection. But, since **the transcendentals don't even exist here on earth, why do we seek them?** It makes no sense for us to seek that which is unattainable. We only seek that which is attainable, if not here then in the hereafter.

What we seek is something beyond our world and beyond our earthly experience. What we seek is God, Who **is** the Perfect Knowledge/Truth, Perfect Love, Perfect Justice/Goodness, Perfect Beauty, and Perfect Home/ Being. For as St. Augustine of Hippo wrote, "Thou hast made us for Thyself and our hearts are restless until they rest in Thee."[1]

Our desire for perfect knowledge, love, justice, beauty, and home affirms not only the existence of God, but also the realization that these desires can only be fulfilled by God. Our conscious desire for these five transcendental qualities provides evidence of a higher level of existence beyond our earthly realm and leads us to a spiritual reality beyond our present experience. This philosophical case for God's existence is traditionally called "The Argument from Desire." Let's next take a closer look at what Fr. Spitzer says about each of these five transcendental desires of the human condition.

(1) Desire for Perfect Knowledge/Truth

Even in children, we see a desire for perfect knowledge when they ask "Why is that?" and when given an answer they then ask the further question, "Well, why is that?" It seems this questioning could go on indefinitely, at least until an adult brings it to an end! This process reveals that children (indeed, all of us) recognize the inadequacy of a partial answer, and that true satisfaction will occur only when a complete and perfect understanding has been achieved.

We humans have a seemingly unlimited longing for "more." We desire to know all there is to know about a subject. And, the more we know the more we want to know. We continue asking questions while realizing we do not yet fully understand all that is to be known.

We know our understanding is not complete. If we did not know it was incomplete, we would not keep asking additional questions. It is our awareness that there is more to be known at the very moment when something is known which drives us to additional questioning. We have an innate awareness of the more.

Such a realization gives rise to another "Why"? Why do we continue asking questions every time something is understood, as if we intuitively know that our current knowledge is limited and does not meet our desire to know all that is to be known? In other words, *how can we be aware of something beyond everything we currently understand?*

Even if someone achieves the highest levels of knowledge in a subject, how does this person know their knowledge still does not explain everything about everything? How do we know what qualifies as a complete and perfect explanation and that we don't yet have it?

This intuitive awareness that there is more to be known than what we now know defies a naturalistic explanation. All our knowledge is incomplete and we know it. But, why are we aware that there is more to be known beyond what we currently know?

The best explanation is that our conscious desire for perfect knowledge and complete truth has been written in our human nature by God, and it is the Perfect Knowledge and Perfect Truth Himself that we seek. Where else could our desire for perfect knowledge, especially knowledge that is not necessary for survival, have come from? This "awareness of the more" indicates the presence of the divine essence (God) to human consciousness and grounds the belief in human transcendentality (the presence of our soul).[2]

(2) Desire for Perfect Love

Spitzer addresses the second transcendental by noting we humans also have a desire for perfect and unconditional love. He states,

> Not only do we have the power to love (i.e., the power to be naturally connected to another human being in profound empathy, emotion, care, self-gift, concern, and acceptance), we have a "sense" of what this profound interpersonal connection would be like if it were perfect.[3]

However, this desire can mislead us into expecting perfect love from another human being. When the relationship does not fulfill our desire for perfect love, this expectation leads to frustration and possibly to a decline in the relationship.

For example, as the imperfections in the love of our beloved manifest themselves (e.g., our spouse is not **perfectly** understanding, kind, forgiving, self-giving, and concerned for me and all my interests), we at first become irritated. This irritation often leads to frustration, which in turn becomes dashed expectations. These dashed expectations may get expressed in either quiet hurt or overt demands, both aimed at trying to achieve a more perfect love from our beloved. When this more perfect love does not happen, thoughts of terminating the relationship may arise. Spitzer notes,

> The root problem was not with the authenticity of this couple's love for one another. It did not arise

> out of a lack of concern, care, and responsiveness
> . . . Rather, it arose out of a false expectation that
> they could be *perfect* and *unconditional* love, truth,
> goodness, fairness, meaning, and home for one
> another.[4]

Why do we fall prey to such an obvious error? Because our desire is for love to be perfect and unconditional, yet the reality is otherwise. We human beings cannot satisfy each other's desire for perfect and unconditional love, no matter how hard we try. Thus, our dissatisfaction and frustration arise from a desire for unconditional and perfect love that has been neither fully experienced nor actualized.

But, what is the origin of this deep desire and yearning for perfect love? Why would we have this desire for perfect love, especially as it just leaves us feeling dissatisfied and frustrated when we cannot find it with another person? ***Why do we have an awareness of and desire for a type of love that we have neither known nor will experience from another human being?***

It seems we are searching for perfect love in all the wrong places. Our desire for perfect and unconditional love can only be met by the Perfect Love (God). Once more, we find that God has implanted in each of us a conscious desire for a perfect love that only God can fulfill. Thus, we see how this second transcendental also provides evidence of God's existence within ourselves.

(3) Desire for Perfect Justice/Goodness

Along with our innate desire for perfect truth/knowledge and perfect love, we also have a conscious desire for perfect justice and ideal goodness. This desire relates to our human sense of good and evil, our capacity for morality, guilt, and nobility, as well as our awareness of what perfect justice and goodness would be like.

For example, even as young children an imperfect manifestation of justice from our parents will elicit the immediate response, "That's not fair!" Adults do the same thing. We feel outrage toward groups, social structures, and even God when we perceive that we have not been treated fairly in life. We truly expect that perfect justice ought to happen, and when it doesn't we feel betrayed.

We only need look at the daily news to find a host of well-meaning, dedicated, and generous people who have tried to obtain perfect justice from the legal system. This can be found in public defenders who decry the legal system for wrongly prosecuting the innocent and victims who vilify the very same system for letting the guilty go free. We seem to expect more justice and goodness than our finite world can deliver, and this causes outrage and even cynicism when it does not come to pass.

What could be the source of our conscious desire for perfect justice and goodness, especially when it seems to be well beyond the actual justice and goodness that we can possibly experience? As Spitzer writes,

> Given that our desire for justice/goodness will only
> be satisfied when we reach perfect, unconditional

Justice/Goodness, it would seem that our desire is guided by a notional awareness of perfect, unconditional Justice/Goodness; and, given that such a notion cannot be obtained from a conditioned and imperfect world, it would seem that its origin is from perfect, unconditional Justice/Goodness itself. For this reason, philosophers have associated this notion of perfect, unconditional Justice/Goodness with the presence of God to human consciousness.[5]

(4) Desire for Perfect Beauty

In western culture, we idolize beauty now more than ever before. Whether it be the Miss Universe pageant, American Idol, or the Biggest Loser, somebody is always looking for or promoting what is conceived as some form of the highest beauty. Wherever we look we believe it can always be better, more beautiful. We also are continually self-critiquing, "I do not look good enough." Spitzer comments,

> Once in a great while, we think we have arrived at consummate beauty. This might occur while looking at a scene of natural beauty: a sunset over the water, majestic green and brown mountains against a horizon of blue sky; but even there, despite our desire to elevate it to the quasi-divine, we get bored and strive for a different or an even more perfect manifestation of natural beauty—a little better sunset, another vantage point of the Alps that's a little more perfect.[6]

As with the other transcendentals, we seem to have an innate awareness of what is most beautiful. This incites us to desire a perfectly beautiful ideal, which leads to both positive and negative results.

The positive result is the continuous human striving for artistic, musical, and literary perfection. We do not desire to just create; we passionately desire to create in ever more beautiful forms the perfection of beauty that we seem to carry within our consciousness. This striving has left a magnificent cultural legacy of architecture, art, music, drama, etc.

However, the negative effect is that we grow bored or frustrated with any imperfect manifestation of beauty. This causes us to try to make perfectly beautiful what is imperfect by nature. For example, a flowering garden can achieve a certain degree of beauty. But, our continued desire to improve it only makes us grow dissatisfied when we cannot perfect it indefinitely.

As with the other transcendentals, we are innately aware of and attracted to perfect beauty itself. But, ***where does our conscious sense of perfect beauty, which does not even exist in our world, come from?*** In other words, since our desire for perfect beauty, art, literature, poetry, and music doesn't have any survival benefit, why would we humans have this conscious desire if God does not exist? Spitzer again concludes,

> Since it seems that the notion of perfect Beauty cannot be obtained or abstracted from a world of sensorial (imperfect) beauty, or even from the beauty of great ideals, goods, and truths (because

they too are conditional and imperfect), one is led to the conjecture that its origin arises out of perfect Beauty itself. For this reason, philosophers have associated this notion of perfect beauty with the notional presence of perfect Beauty (i.e., God) to human consciousness.[7]

(5) Desire for Perfect Home/Being

The fifth and final transcendental is our desire for perfect harmony and peace in our being and our yearning for a perfect experience of home in our world. Spitzer states,

> Human beings also seek a perfect sense of harmony with all that is. They not only want to be at home in a particular environment, they want to be at home with the totality, at home in the cosmos. Have you ever felt, either as a child or an adult, a sense of alienation or discord—a deep sense of not belonging? You ask yourself, "What could be the source?" and you look around and see that at this particular time you have a good relationship with your friends and your family. Your work relationships seem to be going fairly well; community involvements have produced some interesting friends and contexts in which you work. Yet, something's missing. You don't quite feel at home in a *general* sense. Yet you do feel at home with family, friends, organization, etc. You feel like you are out of kilter with, and don't belong to, the *totality*. And yet, all the *specific* contexts you look at seem just fine. You feel emptiness,

a lack of peace, yet there is absolutely nothing you can put your finger on.[8]

When our desire for perfect home is even partially fulfilled, theologians, saints, and mystics throughout the ages have referred to this as joy, love, awe, unity, holiness, and/or peace. For example, in his book, *Surprised by Joy*, C. S. Lewis tried to describe the transcendent joy he experienced. He felt it was the kind of joy that takes over a person and adds a new intensity, awareness, and significance to life.

We again need to ask, **what gives rise to our conscious desire for perfect harmony and our yearning to feel fully at home in our world?** As Spitzer writes,

> It would seem to be linked to perfect Home, perfect Peace, or perfect Harmony itself; for our perception of incompleteness in every concrete manifestation of home reveals that we anticipate more home than any concrete manifestation can deliver; and this, in turn, reveals that we have a notional awareness of perfect home that would not seem to be derivable or abstractable from any concrete experience of home. Thus, the origin of this notional awareness would seem to be traceable to "perfect Home" itself. For this reason, philosophers and theologians have associated it with the presence of God to human consciousness.[9]

In summary, we find evidence of God's existence within human consciousness in our innate awareness of and desire for perfect knowledge/truth, the ideal of perfect love, the idea of perfect justice/goodness, the quality of perfect beauty, and the invitation to perfect home. As none of these transcendental desires can be fulfilled in this world, we have an emptiness that only God can fill. For as C. S. Lewis stated in his book *Mere Christianity,*

> Creatures are not born with desires unless satisfaction for those desires exists. A baby feels hunger: well, there is such a thing as food. . . . If I find in myself a desire which no experience in this world can satisfy, the most probable explanation is that I was made for another world.[10]

World-renowned physicist Sir Arthur Eddington addressed our desire for the transcendent when he wrote,

> We all know that there are regions of the human spirit untrammeled by the world of physics. In the mystic sense of the creation around us, in the expression of art, in a yearning towards God, the soul grows upward and finds the fulfillment of something implanted in its nature. The sanction for this development is within us, a striving born with our consciousness or an Inner Light proceeding from a greater power than ours. Science can scarcely question this sanction, for the pursuit of science springs from a striving that the mind is impelled to follow, a questioning that will not be suppressed. Whether in the intellectual pursuits of science or in the mystical pursuits of the spirit, the light beckons ahead and the purpose surging in our nature responds.[11]

God is perfect and wants us to be one with Him. Thus, our inner craving for perfection must come from and is directed towards God alone. As the *Catechism of the Catholic Church* (#27) states,

> The desire for God is written in the human heart, because man is created by God and for God; and God never ceases to draw man to himself. Only in God will he find the truth and happiness he never stops searching for.

CHAPTER 11

THE CALL OF CONSCIENCE

With our enhanced human consciousness and self-awareness also comes a free will. Free will is the freedom to choose our actions. Humans have the ability to think things through and deliberately decide what to do and how to do it. Our decisions and actions are not based solely on instinct.

Since humans have a free will, shouldn't we be able to act freely and without constraint? But, even though we have the conscious freedom to choose our actions, we don't always feel like we can do whatever we please. What we do needs to make sense to us, it has to feel "right" on some basic level.

Human beings intuitively know there is a difference between right and wrong. We call this innate knowledge of right and wrong our conscience. Moreover, our conscience seems to be trying to guide our actions and get us to behave in a certain way. We feel an inner urge to do right and avoid wrong. We feel an obligation to behave morally, to follow an inner moral law. And, we feel peace when we follow the dictates of our conscience. However, we feel guilty or ashamed whenever we deliberately do something that we know is wrong, even if it is done in secret.

Quite remarkably, this knowledge of right and wrong appears to be universal among mankind. Very few people deny the existence of a conscience, including its guiding influence over our actions. Even if different people's consciences tell them to do or avoid different things, there remains one obligation we all feel: never disobey our conscience. And, while some may try, this inner moral voice cannot be entirely silenced. It continues to try to make itself heard.

Furthermore, our conscience is not necessary for survival. Animals survive just fine without knowing right from wrong. It seems the moral law applies particularly to human beings. Though animals may at times seem to show something of a moral sense, this is certainly not common and the behavior of many animals (particularly predators) is in harsh contrast to any sense of morality.[1]

As noted by geneticist Gerard Verschuuren,

> Morality tells us what ought to be done—no matter what, whether we like it or not, whether we feel it or not, or whether others enforce it or not. Animals, however, live in a world of "what is," not of "what ought to be." They can just follow whatever pops up in their brains. The relationship between predator and prey, for instance, has nothing to do with morality. If predators really had to act morally, their lives would be pretty tough. Animals never do awful things out of meanness or cruelty,

for the simple reason that they have no morality—and thus no cruelty or meanness. But humans
definitely do have the capacity of performing real
atrocities. On the other hand, if animals do seem
to do awful things, it's only because we as human
beings consider their actions "awful" according to
our own standards of morality.[2]

We know humans are different from animals. Unlike
animals, we are guided by an inner compass, a realization
of right and wrong, as well as an awareness of the consequences of our actions. If we do something wrong, we
feel our conscience weighing heavily on our minds and
we feel guilt. As far as we know, animals do not have this
ability to discern between right and wrong.

This brings us to an interesting question: Do humans
know the difference between right and wrong because it
is cultural or because it is inherent within our nature as
human beings? If our knowledge of right and wrong was
entirely cultural, societies would frequently hold differing views of morality. But, we find a surprisingly widespread moral agreement throughout human history.

For example, human societies and cultures throughout history have not held and taught their young that
justice, courage, honesty, kindness, charity, and care
for others who are less fortunate are vices. And, human
societies and cultures have not held and taught their
young that murder, rape, treachery, lying, stealing, and
cheating are virtues. This widespread consensus points
to an underlying basis for this moral agreement that is
inherent in our nature as human beings.

Furthermore, after a series of experiments over eight years, researchers at the Infant Cognition Center at Yale University found that "babies are in fact born with an innate sense of morality, and while parents and society can help develop a belief system in babies, they don't create one."[3] These experiments included showing babies examples of good and bad behavior and observing which behavior the baby preferred. Babies preferred the good behavior 80 percent of the time. Among three-month-old babies, this preference rose to 87 percent.

Paul Bloom, author of *Just Babies: The Origins of Good and Evil* and professor of psychology at Yale, says these experiments show babies are born with a "rudimentary sense of justice" that allows them to judge good and bad in the actions of others. Even though this sense of justice is "tragically limited," the fact that babies can grasp the concepts of good and bad behavior long before they can even speak says a lot about human nature.

These experiments demonstrate that the urging of an inner moral law is indeed "written on our hearts" and is evident even before we have been taught anything about morality. This aspect of human nature poses serious problems for the naturalistic view that humans are nothing more than highly-evolved animals whose instincts were selected for survival value. This view holds that our human tendency to identify right and wrong can be traced to some kind of instinctual evolution, such as that seen when animals protect their young or commonly display warning cries to "protect the herd." However, our conscience operates at a deeper level than our instinctual desires for mere self-preservation and propagation of the species. This is evident when we feel an inner urge to help a stranger in need, even when there is no likelihood of personal gain or benefit. As Francis Collins writes,

> Agape, or selfless altruism, presents a major challenge for the evolutionist.... It cannot be accounted for by the drive of individual selfish genes to perpetuate themselves.[4]

So then, why do people all over the world intuitively know the difference between right and wrong? The most compelling reason is that we were created to be moral beings. We intuitively know the difference between right and wrong and feel called to use our free will accordingly. This suggests that humans, unlike animals, are going to be held accountable for our actions as guided by an informed conscience.

Even though humans are guided by our conscience, and the laws and customs based on that inherent understanding, we are free to do otherwise and often do. Even with our inherent sense of right and wrong, the all-too-common choice of our free will is to do anyway what we feel deep down is not right.

This choice of wrong (usually selfish) actions by people throughout history has resulted in the many flaws and problems we find in human societies today. This is the answer to the question raised in the introduction to Part III of this book; namely, "If God created such a wondrous and perfect universe (as discussed in Part I), why is human society so flawed?" The Church attributes our flawed human society to the cumulative effect of "original sin" (see CCC #408).

This brings us to a very important question: **Why do we humans have a conscience and where did it come from?** How could the source of the moral obligation we feel be something less than our humanity, such as our genes or even our past evolutionary and cultural history?

In *Mere Christianity*, C. S. Lewis eloquently states how our conscience (what he calls the "Moral Law") points to the existence of God,

> If there was a controlling power outside the universe, it could not show itself to us as one of the facts inside the universe—no more than the architect of a house could actually be a wall or staircase or fireplace in that house. The only way in which we could expect it to show itself would be inside ourselves as an influence or a command trying to get us to behave in a certain way. And, that is just what we do find inside ourselves. Surely this ought to arouse our suspicions?[5]

In other words, this very central concept of the knowledge of right and wrong within ourselves is evidence of a God who cares about us. Our conscience is one place where we find evidence of God and also learn about His character. C. S. Lewis goes on to state,

> I find that I do not exist on my own, that I am under a law; that somebody or something wants me to behave in a certain way. . . . Something is directing the universe and it appears in me as a law urging me to do right and making me feel responsible and uncomfortable when I do wrong. . . .

> In the Moral Law somebody or something from
> beyond the material universe is actually getting at
> us. . . . God [is the] impersonal mind at the back of
> the Moral Law.[6]

The fact that all human civilizations and cultures share the ability to differentiate between right and wrong and are guided by an inner moral law, despite their differing geographic locations, traditions, beliefs, and customs, points to God as the source of this moral law. Since this "Moral Law" imprinted on our conscience exists, there must be a moral law-giver. For, there can be no law unless there is a law-giver. That moral law-giver is God. God is the source and ground of the moral obligation we feel deep inside to obey our conscience.

Our conscience provides compelling evidence of a moral law-giver who desires us to be good as He is good. Our conscience has been called "the voice of God in our soul" and it points to a Creator who is infinitely good and loving, as well as perfectly just. In other words, God has instilled a special glimpse of Himself within each one of us.

Our conscience also indicates that God not only created the universe and set it in motion, but also takes an ongoing interest in the individual lives and personal moral choices of each and every human being. Our conscience reveals God cares about each one of us and wants us to be morally good and righteous, just as He is.

As stated by the world's Catholic bishops at the Second Vatican Council,

> In the depths of his conscience, man detects a law

which he does not impose upon himself, but which holds him to obedience. Always summoning him to love good and avoid evil, the voice of conscience when necessary speaks to his heart: do this, shun that. For man has in his heart a law written by God; to obey it is the very dignity of man; according to it he will be judged. Conscience is the most secret core and sanctuary of a man. There he is alone with God, whose voice echoes in his depths.[7]

Of course, we do not always hear that voice clearly. Our conscience can be misread. That is why we have an obligation to properly form our conscience, especially by seeking the truth God has revealed to us in Scripture and the teachings of the Church.

Let's close by looking at how human reason as expressed in philosophy also provides evidence of God's existence.

CHAPTER 12

THE LIGHT OF REASON

There are many metaphysical books and philosophical arguments attempting to logically prove or disprove God's existence. However, the paths often lead away from objective reasoning and rely heavily on subjective interpretation. Consider personal testimonies of miracles as being at one end of the spectrum, as they rely heavily on subjective experience. If you happen to experience one of these miracles, it may be all the evidence you will ever need to believe in God and it may even transform your life, but it may not be very convincing to your agnostic neighbor.

At the other end of the spectrum are objective philosophical arguments, which use logic and reason to make the case for God's existence. In Part I, we looked at a philosophical argument with the Kalam Cosmological Argument, which takes a philosophical position (whatever has a beginning must have a cause) and applies it to our scientific understanding of the origin of the universe.

First, we will discuss two often-cited pragmatic reasons for believing in God. These do not attempt to prove the existence of God, but instead illustrate why it is indeed reasonable and practical to believe in God. We will conclude with Saint Thomas Aquinas's five **philosophical** proofs for God's existence.

Common Consent

The common consent argument is very simple and is probably as old as human history. In a very simplified form, it is essentially the following: Throughout history, the overwhelming vast majority of people, in all places throughout the world, have believed in God or gods, so therefore this is the most reasonable position to hold. Whether it is ancient Egyptians, Mayans, Aborigines, Eskimos, Greeks, Romans, Native Americans, or even today in our increasingly secular world—the vast majority of people believe in at least one God. How many other concepts or ideas have had near unanimous agreement in all corners of the earth and throughout recorded history?

To put this in perspective, consider that an atheist would have to believe that almost all of the people who have ever lived on earth are delusional. It is a position that depends on believing one has the privileged perspective, that one's personal understanding is better than that of nearly all the rest of the human population.

This is not proof by any means, but it does demonstrate that it is a reasonable and common position to believe in God. This should cause an atheist to pause and examine his or her motive for jumping to an opposite conclusion, especially as one cannot conclusively prove that God does not exist.

Pascal's Wager

The second practical reason to believe in God is quite popular and is attributed to the philosopher, mathematician, and physicist Blaise Pascal. In short, it attempts to illustrate that it is more reasonable to live one's life as though you believe in God because to do otherwise is foolish. The argument is based on the following four alternate scenarios:

1. God exists and you believe in Him.
2. God exists but you do not believe in Him.
3. God does not exist but you still believe in Him.
4. God does not exist and you do not believe in Him.

In the last scenario, you may be correct but you will never know you are right because when you die you will become nothing. Likewise in the third scenario, you will never know you were wrong because you will also become nothing after you die.

Moreover, in the first scenario you can gain God's favor and potentially experience eternal happiness in heaven. But in the second scenario, when you die you will find out you are wrong and could face eternal damnation.

In the first scenario you have everything to gain and

in the second scenario you have everything to lose. However, in scenarios three and four you have nothing to gain or lose upon your death because you will never know.

So the only possible way for you to "win the bet" is by believing in God (#1) and the only possible way to lose is by not believing in Him (#2). Therefore, it is most reasonable to believe in God.

This so-called "Pascal's Wager" does not prove anything about the existence of God. But, it should cause an atheist or agnostic to reconsider the odds for success and encourage one to further seek God with a sincere heart.

The "Five Ways" of Saint Thomas Aquinas

In his classical philosophical arguments entitled the "Five Ways," St. Thomas Aquinas explained that God's existence may be known by human reason. However, these "Five Ways" philosophical arguments for God's existence hold validity only if they are clearly understood, which may be difficult in today's fast-paced world. The following are simplified explanations of the "Five Ways" appropriate for today's audience. For a more extensive treatment, see St. Thomas Aquinas' original work, *Summa Theologica* (First Part, Question 2, "The Existence of God").

First Way: First Mover or Argument from Motion

All around us we see that things change and that change is caused by something. As for the thing that changes, although it can be what it will become, it is not yet what it will become.

For example, a lump of black carbon is potentially a brilliant diamond. It actually exists right now in this state (black carbon) and under the right conditions (high temperature and pressure) it can become another state (diamond). Aquinas calls "motion" the change from potentiality to the actual state of having achieved what a thing is capable of (e.g., movement from a lump of black carbon to a diamond).

To explain a change, can we consider only the changing thing? Or, are other factors also involved (e.g., high heat and temperature to bond carbon atoms together to form crystals)? The changing thing begins with only the potential to change, but it needs to be acted upon (moved) by forces outside itself if that potential is to be made actual.

Therefore, Aquinas surmises that whatever is changing cannot do so on its own. If something is changing, it is being changed by another thing, and this by another, and so on. In other words, whatever is in motion (changing) must be put in motion by something else. Nothing sets itself in motion. However, these changes cannot go back infinitely.

Aquinas proposes that the origin of all movement (change) has to begin somewhere. There must be a "First Mover" that is unchanging. Aristotle called this the "Unmoved Mover." And, so we have to come to the first cause of changing that is not itself changed by anything else. We call "God" this unchanging source of change.

Second Way: First Cause or Argument from Cause

Studying "cause and effect" helps us to arrive at Aquinas' second philosophical argument for God's existence. The First Way is an argument from "effect," while the Second Way is an argument from "cause." An "efficient cause" is the immediate preceding source or cause of some effect or movement. For example, if we observe a brilliant diamond, we know this is the effect (result) of a sequence of causes. In other words, the "efficient cause" of the finished diamond is a skilled craftsman cutting and polishing diamond crystals. And, the "efficient cause" of diamond crystals is high heat and temperature bonding carbon atoms to form crystals. And, formation of crystals is only possible after something caused carbon atoms to form. And, so on. The finished diamond could not have caused itself. Rather, it is the resultant effect of a long sequence of efficient causes.

Therefore, similar to our First Mover argument the First Cause argument states that nothing causes itself to be. Ultimately, everything in existence has been caused by something else. As before, this chain of causes cannot extend backward indefinitely. There must be something uncaused, something on which all things that need a cause of their existence are dependent. There must have been an Uncaused First Cause, which we call God.

Third Way: Necessary Being or Argument from Possibility and Necessity

Aquinas begins his third argument by considering that all around us things come into being and go out of being.

For example, a plant grows from a tiny shoot, flowers brilliantly, then withers and dies.

Therefore, whatever comes into being or goes out of being does not have to be. It is possible but not necessary, as nonexistence is a real possibility. Thus, a "possible being" is something that does not always exist, but is possible to be (to exist) and not to be (to not exist). Rather, Aquinas calls something that always exists and must exist a "necessary being."

Since "possible beings," like us and our universe, have not always existed and if everything that exists is a "possible being," then at one time there must have been nothing. And if this were so, then nothing would now exist, because only nothing comes from nothing. However, we are something where nothing should be.

Therefore, not everything that exists is a "possible being." "Possible beings" (like us and our universe) could never have come to exist, unless there is a being who is necessary—who must exist, who by necessity has always existed. That Necessary Being is God.

Fourth Way: Greatest Being or Argument from Degrees of Perfection

We notice around us things that vary in degrees. For example, one diamond can be lighter or darker than another. A freshly baked pie is hotter than one taken out of the oven hours before. Thus, among things there is a gradation of "more" and "less." Some things are more good, more perfect, more beautiful, and other things less so.

When we arrange things in terms of "more" and "less," we naturally think of them on a scale approaching "most" and "least." For example, we think of a lighter color as approaching the brightness of pure white and a darker color as approaching the opacity of pitch black. This means that we think of them at various "distances" from the extremes and as possessing, in degrees of "more" or "less," what the extremes possess in full measure. In other words, such "more" or "less" comparisons move towards an end condition and can be graded in increments as they approach their ultimate condition. For instance, the color of diamonds is better the closer they approach what is pure white.

Thus, since degrees of perfection exist there must also exist a "best," a source and real ultimate standard for comparison of all the varying degrees of perfection we recognize. There is, therefore, an ultimately best and top state beyond compare that is the "most" complete and perfect. In other words, to recognize degrees of perfection there must be an ultimate perfection; namely, someone or something who is absolutely perfect—and that Greatest Being is God.

Fifth Way: Supreme Ordering Intelligence or Argument from Governance of the World

The fifth philosophical argument Aquinas proposes is based on the regularity and order that exists throughout nature. For example, the sun rises and sets every day and the four seasons are always in sequential order and produce the same basic results year after year.

This order exists in spite of the fact that the affected bodies (e.g., sun, planets) are without awareness and lack intelligence. The regular and orderly behavior of objects that lack awareness and intelligence cannot be the result of chance. Anything without awareness tends to behave in an orderly and intelligent manner only under intelligent guidance. If there was no intelligent and governing agent to provide order, then the result must be without order; in other words, there would be chaos. Instead, since there is order and regularity in the universe, it must be the result of governance from some absolutely intelligent being. That Supreme Ordering Intelligence is God, who provides order and harmony in the universe through the governing laws of nature that He created.

CONCLUSION TO PART III

In the first two parts of this book we examined scientific evidence of God's existence in the cosmos. This evidence includes the Big Bang, the mathematically elegant and intelligible laws of nature, the amazingly precise fine tuning of the universe and our planet for life, and the biological evidence of an intelligent cause for the informational content of DNA.

In this part, we looked at inner evidence (within human nature) for the existence of God. This includes human consciousness, with our desire for perfect knowledge, love, justice, beauty, and home, which can be found only in God, and our conscience—the "Moral Law" within us that serves to guide the actions of our free will.

We can call this evidence the "3 Cs" (cosmos, consciousness, and conscience). And, we've seen how these "3 Cs", as well as the "Five Ways" philosophical reasoning, provide compelling evidence for the existence of God.

Christians believe that evidence of God's existence can also be found in God's Revelations of Himself to humankind. Christians believe that God has revealed Himself to us in three ways, in God's three "Spoken Words."[1]

First, God's *creative word* brought the universe into existence at the creation. As discussed in Part I, science can lead us to a deeper understanding of God's creative word by discovering the orderly laws and principles that govern the cosmos, which has been called the "Book of God's Works." Second, God has given us His *written word* in the Bible, which is the "Book of God's Words." And, third, God has also revealed Himself in the *Living Word*, Jesus Christ. Christians believe that Jesus <u>C</u>hrist is the "4th C" supporting God's existence, in addition to the <u>c</u>osmos, our <u>c</u>onsciousness, and our <u>c</u>onscience.

While the "3 Cs" and God's three "Spoken Words" provide compelling and rational evidence for the existence of God, this evidence is *not* proof in a scientific and material sense and never can be. Yet, it is perfectly normal to wonder why God doesn't more clearly reveal Himself and unquestionably prove His existence to us.

It doesn't take faith to accept the existence of the sun or the moon. They are plain to see. However, *God fully respects our free will. He has chosen to offer us an invitation that isn't overshadowed by coercion, such as would occur with an overpowering revelation of Himself that would leave us cowering in submission.* Rather, God has given us the freedom to choose to believe in His existence—or not.

This is articulated well by Norman Geisler in his book entitled, *I Don't Have Enough Faith to Be an Atheist,*

> God has provided each of us with the opportunity
> to make an eternal choice to either accept him or
> reject him. And in order to ensure that our choice

is totally free, he puts us in an environment that is filled with evidence of his existence, but without his direct presence—a presence so powerful that it could overwhelm our freedom and thus negate our ability to reject him. In other words, ***God has provided enough evidence in life to convince anyone willing to believe, yet has also left some ambiguity so as to not compel the unwilling.*** In this way, God gives us the opportunity to either love him or reject him without violating our freedom. For love, by definition, must be freely given. It cannot be coerced.[2]

While there is compelling evidence (e.g., the "3 Cs"), it is not possible to absolutely prove the existence of God beyond any doubt. A "leap of faith" is required from each of us. It is ultimately our choice as to whether or not we will believe in God and seek to live our lives in accordance with God's Will. Since true love cannot be forced, God desires that we each *freely choose* to know, love, and serve Him. It is our sincere hope and prayer that this book will help you do that and be able to better model this lifestyle for others.

PART III:
PERSONAL REFLECTION/
DISCUSSION QUESTIONS

1. If God created such a wondrous and perfect universe, why is human society so flawed?

2. What are the five transcendentals and do you agree that they provide evidence of God's existence? Why or why not?

3. Do you agree that our knowledge of right and wrong provides evidence for God's existence? Why or why not?

4. Do you agree that it is not possible to conclusively prove without a doubt that God exists? Why or why not?

5. How could you help someone who is intellectually open to the possibility of God's existence make the "leap of faith" to a personally committed faith? What are the next steps to an active and committed faith?

SELECT BIBLIOGRAPHY

For Part I: Cosmic Evidence of God's Existence

DVDs:

Cosmic Origins, Copyright 2012 Ignatius Press (www. ignatius.com)

The Privileged Planet, Copyright 2004 Illustra Media (www.illustramedia.com)

The Case for a Creator, Copyright 2006 Illustra Media (www.illustramedia.com)

The Evidence: God, the Universe and Everything, Copyright 2002 Faith for Today (www.theevidence.tv)

The Universe: God and the Universe, Copyright 2011 A&E Networks (www.history.com)

Books:

Stephen H. Barr, *Modern Physics and Ancient Faith* (Notre Dame, Indiana: University of Notre Dame Press, 2006). See especially Part II "In the Beginning," Chapter 11 "The Design Argument and the Laws of Nature," and Chapter 15 "The Anthropic Coincidences."

Jimmy H. Davis and Harry L. Poe, *Designer Universe: Intelligent Design and the Existence of God* (Nashville, Tennessee: Broadman & Holman Publishers, 2002). See especially Chapter Four, "The Universe: Designer Showcase."

Patrick Glynn, *God: The Evidence: The Reconciliation of Faith and Reason in a Postsecular World* (Rocklin, California: Prima Publishing, 1999). See especially Chapter One, "A Not-So-Random Universe."

Guillermo Gonzales and Jay W. Richards, *The Privileged Planet: How Our Place in the Cosmos Is Designed for Discovery* (Washington, D.C.: Regnery, 2004).

Robert J. Spitzer, S.J., *New Proofs for the Existence of God: Contributions of Contemporary Physics and Philosophy* (Grand Rapids, Michigan: Wm. B. Eerdmans Publishing Company, 2010). See especially Part One, "Indications of Creation and Supernatural Design in Contemporary Big Bang Cosmology."

Lee Strobel, *The Case for a Creator* (Grand Rapids, Michigan: Zondervan, 2004). See especially Chapter 5, "The Evidence from Cosmology: Beginning with a Bang," and Chapter 6, "The Evidence of Physics: The Cosmos on a Razor's Edge."

For Part II: Biological Evidence of God's Existence

DVD:
Unlocking the Mystery of Life, Copyright 2002 Illustra Media (www.illustramedia.com)

Books:
Pope Benedict XVI, *In the Beginning: A Catholic Understanding of the Story of Creation and the Fall* (Grand Rapids, Michigan: William B. Eerdmans Publishing Co., 1995).

Pope Benedict XVI, Christoph Cardinal Schönborn, et.al., *Creation and Evolution: A Conference with Pope Benedict XVI* (San Francisco: Ignatius Press, 2008).

Christopher T. Baglow, *Faith, Science and Reason: Theology on the Cutting Edge* (Woodbridge, Illinois: Midwest Theological Institute, 2009). See especially Chapter Eight, "Going 'Deeper than Darwin': Faith and the Origin of Living Things."

Francis S. Collins, *The Language of God: A Scientist Presents Evidence for Belief* (New York: Free Press, 2006).

Jimmy H. Davis and Harry L. Poe, *Designer Universe: Intelligent Design and the Existence of God* (Nashville, Tennessee: Broadman & Holman Publishers, 2002). See especially Chapter Six, "Designer Genes."

William A. Dembski and James M. Kushiner, *Signs of Intelligence* (Grand Rapids, Michigan: Brazos Press,

2001). See especially Chapter 8, "Word Games: DNA, Design, and Intelligence."

Karl W. Giberson and Francis S. Collins, *The Language of Science and Faith: Straight Answers to Genuine Questions* (Downers Grove, Illinois: InterVarsity Press, 2011).

John F. Haught, *Making Sense of Evolution: Darwin, God, and the Drama of Life* (Louisville, Kentucky: Westminster John Knox Press, 2010).

George Sim Johnston, *Did Darwin Get It Right? Catholics and the Theory of Evolution* (Huntington, Indiana: Our Sunday Visitor, 1998).

Denis O. Lamoureux, *Evolutionary Creation: A Christian Approach to Evolution* (Eugene, Oregon: Wipf and Stock Publishers, 2008).

Christoph Cardinal Schönborn, *Chance or Purpose? Creation, Evolution, and a Rational Faith* (San Francisco: Ignatius Press, 2007).

Gerald M. Verschuuren, *God and Evolution? Science Meets Faith* (Boston: Pauline Books & Media, 2012).

For Part III: Human Evidence of God's Existence

DVD:

The Question of God: Sigmund Freud & C. S. Lewis, Copyright 2004 Tatge-Lasseur Productions, distributed by PBS Home Video, a department of the Public Broadcasting Service (PBS) (www.pbs.org)

Books:

Francis S. Collins, *The Language of God: A Scientist Presents Evidence for Belief* (New York: Free Press, 2006). See especially Chapter One, "From Atheism to Belief."

Norman L. Geisler and Frank Turek, *I Don't Have Enough Faith to be an Atheist* (Wheaton, Illinois: Crossway Books, 2004).

Peter Kreeft and Ronald K. Tacelli, S.J., *Handbook of Catholic Apologetics: Reasoned Answers to Questions of Faith* (San Francisco: Ignatius Press, 2009).

C. S. Lewis, *Mere Christianity* (New York: HarperCollins Publishers, 2001). See especially Book One, "Right and Wrong as a Clue to the Meaning of the Universe."

Robert J. Spitzer, S.J., *New Proofs for the Existence of God: Contributions of Contemporary Physics and Philosophy* (Grand Rapids, Michigan: Wm. B. Eerdmans Publishing Company, 2010). See especially Chapter Eight, "The Human Mystery: Five Yearnings for the Ultimate."

NOTES

Preface

1. See: http://www.pewforum.org/2015/05/12/americas-changing-religious-landscape/.

2. Ibid.

3. See: http://www.stephenjaygould.org/ctrl/news/file002.html.

4. John H. Westerhoff, III, *Will Our Children Have Faith?* (Harrisburg, Pennsylvania: Morehouse Publishing, 2000), pp.87-103.

5. Ibid., p. 97.

6. See: http://www.catholicnews.com/data/stories/cns/1103264.htm.

7. See: http://www.britannica.com/EBchecked/topic/30078/apologetics.

8. See: http://www.archmil.org/bishops/Pallium-Lecture-Series/Cardinal-Dolan-2013.htm (Part 2).

9. See http://www.vatican.va/holy_father/francesco/apost_exhortations/documents/papa-francesco_esortazione-ap_20131124_evangelii-gaudium_en.html (paragraph #132).

Chapter 1: The Origin of the Beginning

1. See: *The Privileged Planet* DVD (La Mirada, California: Illustra Media, 2004).

2. Arthur Eddington, "The End of the World: From the Standpoint of Mathematical Physics," *Nature*, Vol. 127 (1931), p.450.

3. See: *The Universe: God and the Universe* DVD (A&E Networks, 2011). See also, Robert J. Spitzer, *New Proofs for the Existence of God: Contributions of Contemporary Physics and Philosophy* (Grand Rapids, Michigan: Wm. B. Eerdmans Publishing Company, 2010), pp. 4-5 and Chapter One, "Indications of Creation in Big Bang Cosmology."

4. Stephen H. Barr, *Modern Physics and Ancient Faith* (Notre Dame, Indiana: University of Notre Dame Press, 2006), pp. 58-61.

5. See: *The Case for a Creator* DVD (La Mirada, California: Illustra Media, 2006).

6. Quentin Smith, "The Uncaused Beginning of the Universe," in William Lane Craig and Quentin Smith, *Theism, Atheism, and Big Bang Cosmology* (Oxford: Clarendon, 1993), p. 135.

7. William Lane Craig, *The Kalam Cosmological Argument* (Eugene, Oregon: Wipf & Stock Publishers, 2000).

8. Edward T. Whittaker, *The Beginning and End of the World* (Oxford: Oxford University Press, 1942), quoted in: Robert Jastrow, *God and the Astronomers*, p. 103.

9. Allan Sandage, interview quoted in: Fred Heeren, *Show Me God: What the Message from Space Is Tell-*

ing Us About God (Olathe, Kansas: Day Star Publications, 2004), p. 224.

10. Robert Jastrow, *God and the Astronomers* (New York: W.W. Horton, 1992), pp. 14 & 207.

11. See: Robert J. Spitzer, *New Proofs for the Existence of God: Contributions of Contemporary Physics and Philosophy*, pp. 33-44.

12. Alexander Vilenkin, *Many Worlds in One: The Search for Other Universes* (New York: Hill and Wang, 2006), p. 176.

Chapter 2: The Architect of the Laws of Nature

1. See: *The Universe: God and the Universe* DVD.

2. Barr, *Modern Physics and Ancient Faith*, p. 77.

3. Albert Einstein, *The World As I See It* (New York: Open Road Integrated Media, 2011), p. 37.

4. Quoted in: *The Privileged Planet* DVD.

5. Ibid.

6. Ibid.

7. See: *The Privileged Planet* DVD.

8. Quoted in: *The Privileged Planet* DVD.

9. Quoted in: *The Evidence: God, the Universe and Everything* DVD (Semi Valley, California: Faith for Today, 2002).

10. Paul Davies, "The Christian Perspective of a Scientist," Review of "The Way the World Is" by John Polkinghorne, *New Scientist*, Vol. 98, No. 1354, 2 June 1983, p.638.

11. Quoted in: *The Universe: God and the Universe* DVD.

Chapter 3: The Universal Fine Tuner

1. Barr, *Modern Physics and Ancient Faith*, p. 25.

2. See: http://www.physics.arizona.edu/~haar/ADV_LAB/BIG_G.pdf.

3. See: *The Universe: God and the Universe* DVD

4. Quoted in: *The Privileged Planet* DVD.

5. See: *The Privileged Planet* DVD.

6. Paul Davies, *God and the New Physics* (New York: Penguin Putnam Inc., 1983), p. 188.

7. Jimmy H. Davis and Harry L. Poe, *Designer Universe: Intelligent Design and the Existence of God* (Nashville, Tennessee: Broadman & Holman Publishers, 2002), p. 86.

8. Ibid.

9. Barr, *Modern Physics and Ancient Faith*, p. 125.

10. Barr, *Modern Physics and Ancient Faith*, p. 126.

11. Davies, *God and the New Physics*, p. 187-188.

12. Barr, *Modern Physics and Ancient Faith*, p. 120.

13. Stephen Hawking and Leonard Mlodinow, *The Grand Design* (New York: Bantam Books, 2010), p. 160.

14. William A. Dembski and James M. Kushiner, *Signs of Intelligence* (Grand Rapids, Michigan: Brazos Press, 2001), p 167.

15. Hawking and Mlodinow, *The Grand Design*, p. 160.

16. See: *The Evidence: God, the Universe and Everything* DVD.

17. See: *The Universe: God and the Universe* DVD.

18. Quoted in: *The Universe: God and the Universe* DVD.

19. Quoted in: *The Evidence: God, the Universe and Everything* DVD.

20. Stephen W. Hawking, *A Brief History of Time—From the Big Bang to Black Holes* (New York: Bantam Books, 1988), p. 125.

21. Charles Townes quoted in Bonnie Azab Powell, " 'Explore as much as we can': Nobel Prize winner Charles Townes on evolution, intelligent design, and the meaning of life," *UC Berkley News Center* (June 17, 2005). See: http://www.berkeley.edu/news/ media/releases/2005/06/17_townes.shtml.

22. Patrick Glynn, *God: The Evidence: The Reconciliation of Faith and Reason in a Postsecular World* (Rocklin, California: Prima Publishing, 1999), p. 22.

23. Quoted in: *The Evidence: God, the Universe and Everything* DVD.

24. Francis S. Collins, *The Language of God: A Scientist Presents Evidence for Belief* (New York: Free Press, 2006), p. 77. Reprinted with permission of Simon & Schuster Publishing Group. All rights reserved.

25. Fred Hoyle, "The Universe: Past and Present Reflections," *Annual Reviews of Astronomy and Astrophysics*, Vol. 20 (1982), p. 16.

26. Glynn, *God: The Evidence*, p. 53.

27. Davies, *God and the New Physics*, pp. 173-174.

28. Edward Harrison, *Masks of the Universe: Changing Ideas on the Nature of the Cosmos* (London: Macmillan Publishing Company, 1985), pp. 252 & 263.

Chapter 4: The Local Fine Tuner

1. Peter D. Ward and Donald Brownlee, *Rare Earth: Why Complex Life is Uncommon in the Universe* (New York: Copernicus Books, 2004), p. 193.

2. See: *The Privileged Planet* DVD.

3. Ward and Brownlee, *Rare Earth*, pp. 20-21.

4. See: *The Privileged Planet* DVD.

5. Ibid.

6. Ward and Brownlee, *Rare Earth*, p. 213.

7. Ward and Brownlee, *Rare Earth*, p. 237.

8. Ward and Brownlee, *Rare Earth*, p. 194.

9. Ibid.

10. See: *The Privileged Planet* DVD.

11. Ibid.

12. See: Collins, *The Language of God*, p. 89.

13. See: http://www.scientificamerican.com/article.cfm?id=origin-of-oxygen-in-atmosphere.

14. Ward and Brownlee, *Rare Earth*, p. 98.

15. Ward and Brownlee, *Rare Earth*, pp. 222-226.

16. Ibid.

17. Ward and Brownlee, *Rare Earth*, p. 222.

18. See: *The Privileged Planet* DVD.

19. Ibid.

20. Davis and Poe, *Designer Universe*, p. 100.

21. Ward and Brownlee, *Rare Earth*, p. 238-239.

22. Davis and Poe, *Designer Universe*, p. 102.

23. H. Wayne House, Gen. Ed., *Intelligent Design 101:*

Leading Experts Explain the Key Issues, Chapter 5, Jay W. Richards, "Why Are We Here? Accident or Purpose?" (Grand Rapids, Michigan: Kregel Publications), p. 141.

24. Davis and Poe, *Designer Universe,* p. 99.

25. Ibid.

26. Davis and Poe, *Designer Universe,* p. 100.

27. See: *The Privileged Planet* DVD.

28. Quoted in: *The Privileged Planet* DVD.

29. Ward and Brownlee, *Rare Earth,* pp. 230-234.

30. See: *The Privileged Planet* DVD.

31. Quoted in: *The Privileged Planet* DVD.

32. House, *Intelligent Design 101,* p. 152.

33. See: *The Privileged Planet* DVD.

34. Quoted in: *The Case for a Creator* DVD.

Conclusion to Part I

1. Quoted in: *The Evidence: God, the Universe and Everything* DVD.

2. Quoted in: *The Universe: God and the Universe* DVD

3. Quoted in: *The Universe: God and the Universe* DVD

Introduction to Part II

1. George Gaylord Simpson, *The Meaning of Evolution,* revised edition (New Haven, Connecticut: Yale University Press, 1967), p. 345.

2. Karl W. Giberson and Francis S. Collins, *The Language of Science and Faith: Straight Answers to Genuine Questions* (Downers Grove, Illinois: InterVarsity Press, 2011), p. 114.

Chapter 5: The Truth of Genesis

1. Pope John Paul II, General Audience January 29, 1986, "In Creation God Calls The World Into Existence From Nothingness." (emphasis added) See: http://inters.org/John-Paul-II-Catechesis-Creation-Nothingness.

2. Pope Benedict XVI, Christoph Cardinal Schönborn, et.al., *Creation and Evolution: A Conference with Pope Benedict XVI* (San Francisco: Ignatius Press, 2008), p. 91.

3. George Sim Johnston, *Did Darwin Get It Right? Catholics and the Theory of Evolution* (Huntington, Indiana: Our Sunday Visitor, 1998), p. 120.

4. John Hammond Taylor, tr., *Ancient Christian Writers: St. Augustine, The Literal Meaning of Genesis*, Volume 1 (Mahwah, New Jersey: Paulist Press, 1982), Book One, Chapter 19, pp. 42-43.

Chapter 6: The Evolutionary Creator

1. Denis O. Lamoureux, *Evolutionary Creation: A Christian Approach to Evolution* (Eugene, Oregon: Wipf and Stock Publishers, 2008), pp. 29-35.

2. See: http://www.vatican.va/holy_father/benedict_xvi/speeches/2008/october/documents/hf_ben-xvi_spe_20081031_academy-sciences_en.html.

3. Pope John Paul II, General Audience July 10, 1985, "The Proofs for God's Existence." See: http://inters.org/John-Paul-II-Science-Proofs-God.

4. Pope Benedict XVI, Christoph Cardinal Schönborn, et.al., *Creation and Evolution: A Conference with Pope Benedict XVI*, p. 92.

5. Michael J. Behe, *The Edge of Evolution: The Search for the Limits of Darwinism* (New York: Free Press, 2007), pp. 3, 12, 64-65.

6. Darrel R. Falk, *Coming to Peace with Science: Bridging the Worlds Between Faith and Biology* (Downers Grove, Illinois: InterVarsity Press, 2004), p. 84.

7. Behe, *The Edge of Evolution*, p. 70.

8. John F. Haught, *Making Sense of Evolution: Darwin, God, and the Drama of Life* (Louisville, Kentucky: Westminster John Knox Press, 2010), p. 43.

9. Charles Darwin, *On the Origin of Species: The Illustrated Edition* (New York: Sterling Publishing Company, 2008), pp. 512-513.

10. International Theological Commission (ITC), *Communion and Stewardship: Human Persons Created in the Image of God*, 2004. See paragraph #68 at http://www.vatican.va/roman_curia/congregations/cfaith/cti_documents/rc_con_cfaith_doc_20040723_communion-stewardship_en.html.

11. Falk, *Coming to Peace with Science*, p. 206.

12. Johnston, *Did Darwin Get It Right? Catholics and the Theory of Evolution*, p. 121.

13. John Hammond Taylor, tr., *Ancient Christian Writers: St. Augustine, The Literal Meaning of Genesis*, Volume 1, Book Five, Chapter 7, pp. 158-159.

14. John Hammond Taylor, tr., *Ancient Christian Writers: St. Augustine, The Literal Meaning of Genesis*, Volume 1, Book Five, Chapter 23, pp. 174-175.

15. Johnston, *Did Darwin Get It Right? Catholics and the Theory of Evolution*, p. 121.

16. Pope Benedict XVI, "Homily of His Holiness Benedict XVI," April 24, 2005. See: http://www.vatican.va/holy_father/benedict_xvi/homilies/2005/documents/hf_ben-xvi_hom_20050424_inizio-pontificato_en.html.

Chapter 7: Not Revealed by Evolutionary Naturalism

1. Richard Dawkins, "Man vs. God: Richard Dawkins argues that evolution leaves God with nothing to do," *Wall Street Journal*, September 12, 2009.

2. Francis Crick, *What Mad Pursuit: A Personal View of Scientific Discovery* (New York: Basic Books, 1990), p. 138.

3. Richard Dawkins, *The Blind Watchmaker: Why the Evidence of Evolution Reveals a Universe without Design* (New York: W. W. Norton & Company, 1996), p. 6.

4. Cardinal Christoph Schönborn, "Reasonable Science, Reasonable Faith," *First Things*, April 2007.

5. See: http://www.goodreads.com/quotes/178439-the-cosmos-is-all-that-is-or-was-or-ever.

6. John P. Wiley Jr., quoting cosmologist Edward R. Harrison, *Smithsonian Magazine*, December, 1995.

7. Haught, *Making Sense of Evolution*, p. 18.

8. *Youth Catechism of the Catholic Church* (San Francisco: Ignatius Press, 2010), p. 37.

9. Pope John Paul II, General Audience January 29, 1986, "In Creation God Calls The World Into Existence From Nothingness." See: http://inters.org/John-Paul-II-Catechesis-Creation-Nothingness.

10. Pope John Paul II, Message to the Pontifical Academy of Sciences, "Magisterium Is Concerned with Question of Evolution for It Involves Conception of Man," October 22, 1996, printed in October 30, 1996 issue of English edition of L'Osservatore Romano. See: https://www.ewtn.com/library/PAPALDOC/ JP961022.HTM.

11. Pope John Paul II, General Audience April 16, 1986, "Humans are Spiritual and Corporeal Beings." See: http://inters.org/John-Paul-II-Catechesis-Spiritual-Corporeal.

12. International Theological Commission, *Communion and Stewardship: Human Persons Created in the Image of God*, 2004, paragraph #69.

13. Francis S. Collins, *The Language of God: A Scientist Presents Evidence for Belief* (New York: Free Press, 2006), p. 205. Reprinted with permission of Simon & Schuster Publishing Group. All rights reserved.

14. Pope John Paul II, General Audience March 5, 1986, "Creation is the Work of the Trinity." See: http://inters.org/John-Paul-II-Catechesis-Creation-Trinity.

15. Christoph Cardinal Schönborn, "Finding Design in Nature," *New York Times*, July 7, 2005.

Chapter 8: Found in a Layered Explanation of Evolution

1. Haught, *Making Sense of Evolution*, pp. 23–25.

Chapter 9

1. For a short YouTube video that uses excellent computer graphics to illustrate this process, see: http://youtu.be/CAc9oNjXe0M?t=1m20s.

2. Dembski and Kushiner, *Signs of Intelligence*, p. 107.

3. Ibid.

4. Bill Gates, *The Road Ahead* (Boulder, Colorado: Blue Penguin, 1996), p. 228.

5. Richard Dawkins, *River Out Of Eden: A Darwinian View Of Life* (New York: Basic Books, 1996), p. 10.

6. See: http://www.teachastronomy.com/astropedia/article/Life-as-Digital-Information.

7. Dembski and Kushiner, *Signs of Intelligence*, p. 115.

8. See: http://www.news-medical.net/health/Sickle-Cell-Disease-Genetics.aspx.

9. Henry Quastler, *The Emergence of Biological Organization* (New Haven, Connecticut: Yale University Press, 1964), p. 67.

10. Dembski and Kushiner, *Signs of Intelligence*, p. 116.

11. Anthony Flew, with Roy Abraham Varghese, *There is a God: How the World's Most Notorious Atheist Changed His Mind* (New York: HarperCollins, 2007), p. 127.

12. Stephen C. Meyer, *Signature in the Cell: DNA and the Evidence for Intelligent Design* (New York: HarperCollins Publishers, 2009), p. 347.

13. Francis S. Collins, *The Language of God: A Scientist Presents Evidence for Belief* (New York: Free Press, 2006), pp. 123-124. Reprinted with permission of Simon & Schuster Publishing Group. All rights reserved.

Conclusion to Part II

1. Pope Benedict XVI, *In the Beginning: A Catholic Understanding of the Story of Creation and the Fall* (Grand Rapids, Michigan: William B. Eerdmans Publishing Co., 1995), pp. 56-57.

Chapter 10: The Core of Consciousness

1. St. Augustine, *Confessions* (New York: Oxford University Press, 1991), Book 1, Chapter 1.1.

2. Spitzer, *New Proofs for the Existence of God*, p. 266.

3. Ibid.

4. Ibid., p. 267.

5. Ibid., p. 270.

6. Ibid., p. 277.

7. Ibid., p. 280.

8. Ibid., p. 281.

9. Ibid., p. 282.

10. C.S. Lewis, *Mere Christianity* (New York: Harper-Collins Publishers, 2001), pp. 136-137.

11. Sir Arthur Eddington, *The Nature of the Physical World* (Cambridge: Cambridge University Press, 1928), pp. 327-328.

Chapter 11: The Call of Conscience

1. See: Collins, *The Language of God*, p. 23.

2. Gerard M. Verschuuren, "Morality Is Not a Biological Issue." See: http://www.strangenotions.com/morality-is-not-a-biological-issue/.

3. Susan Chun, "Are we Born with a Moral Core?

The Baby Lab says 'Yes', " See: http://www.cnn.
com/2014/02/12/us/baby-lab-morals-ac360/index.
html.

4. Francis S. Collins, *The Language of God: A Scientist
    Presents Evidence for Belief* (New York: Free Press,
    2006), p. 27. Reprinted with permission of Simon &
    Schuster Publishing Group. All rights reserved.

5. Lewis, *Mere Christianity*, p. 24.

6. Ibid., pp. 25, 28, 32.

7. *Pastoral Constitution on the Church in the Modern
    World* (Gaudium et Spes), Promulgated by Pope
    Paul VI on December 7, 1965, paragraph #16.

Conclusion to Part III

1. See: *Jesus: Fact or Fiction?* DVD (Inspirational Films,
    2003)

2. Norman L. Geisler and Frank Turek, *I Don't Have
    Enough Faith to be an Atheist* (Wheaton, Illinois:
    Crossway Books, 2004), p. 31.

INDEX